PoliTO Springer Series

Springer, in cooperation with Politecnico di Torino, publishes the PoliTO Springer Series. This co-branded series of publications includes works by authors and volume editors mainly affiliated with Politecnico di Torino and covers academic and professional topics in the following areas: Mathematics and Statistics, Chemistry and Physical Sciences, Computer Science, All fields of Engineering. Interdisciplinary contributions combining the above areas are also welcome. The series will consist of lecture notes, research monographs, and briefs. Lectures notes are meant to provide quick information on research advances and may be based e.g. on summer schools or intensive courses on topics of current research, while SpringerBriefs are intended as concise summaries of cutting-edge research and its practical applications. The PoliTO Springer Series will promote international authorship, and addresses a global readership of scholars, students, researchers, professionals and policymakers.

THE SERIES IS INDEXED IN SCOPUS

Maria Giulia Ballatore

Empower Education to Foster Engineering

Springer

POLITECNICO DI TORINO

Maria Giulia Ballatore
Department of Mathematical Sciences
'G.L. Lagrange'
Politecnico di Torino
Torino, Italy

ISSN 2509-6796 ISSN 2509-7024 (electronic)
PoliTO Springer Series
ISBN 978-3-031-75894-2 ISBN 978-3-031-75895-9 (eBook)
https://doi.org/10.1007/978-3-031-75895-9

This Springer imprint is published by the registered company Springer Nature Switzerland AG
The registered company address is: Gewerbestrasse 11, 6330 Cham, Switzerland

To my beloved family

Preface

This book provides a comprehensive exploration of the vital role that Engineering Education Researchers (EER) play within Politecnico di Torino, an Italian Technical University. This monograph is born out of the increasing recognition that bridging the gap between educational theory and practical application is crucial for advancing engineering education.

The journey begins with an introduction to the landscape of EER, setting the stage within the unique Italian context. This foundational framework is essential for understanding the subsequent analyses and discussions.

Central to this work are two pivotal chapters. Chapter 2 delves into the perspectives of students, shedding light on their experiences, challenges, and feedback. Chapter 3 shifts focus to the lecturers, exploring their methodologies, innovations, and the obstacles they encounter in implementing educational innovations. These chapters are rich with case studies, research findings, and project evaluations, offering a balanced view from both sides of the educational process.

One of the key issues addressed is the disconnection often observed between educational theory and practical implementation. This book aims to provide actionable insights and practical recommendations to harmonize teaching methods with educational research findings. By understanding the experiences and challenges of both students and educators, this monograph seeks to foster an environment where educational practices are continuously improved and aligned with real-world demands.

The concluding chapter synthesizes the findings and discussions, offering a comprehensive overview and proposing final recommendations for enhancing engineering education. The goal is not just to highlight problems but to offer concrete solutions and strategies for improvement.

This monograph is intended for educators, researchers, and policymakers dedicated to advancing the quality of engineering education. It aims to serve as a valuable resource for those involved in the academic and professional development of engineering students, providing a holistic view of the educational landscape and practical solutions for its improvement.

Torino, Italy
July, 2024

Maria Giulia Ballatore

Acknowledgments

The development of this book has been a collaborative effort, drawing on the expertise and insights of numerous colleagues and students. I am particularly grateful to those who participated in the successive developments of this subject, whose contributions have been invaluable.

I express my gratitude to the institutions and funding bodies that supported this work, in particular the European Union through the Erasmus+ and Capacity Building projects, as well as the Compagnia di San Paolo and Fondazione Agnelli. Their support has been instrumental in bringing this project to fruition.

I would like to express my deepest gratitude to Prof. Anita Tabacco, my supervisor and the driving force behind this work. Collaborating with her is an immense privilege that goes beyond words. Her unyielding curiosity and continual generation of ideas cultivate an environment that is not only stimulating but also remarkably innovative.

Finally, I extend my thanks to all the colleagues who have supported the TEACH group for their open-mindedness and willingness to pilot educational innovations. Their contributions have been crucial in advancing this work.

Declarations

Competing Interests This study includes projects funded by Compagnia di San Paolo, Fondazione Agnelli and European Union. The authors have no conflicts of interest to declare that are relevant to the content of this chapter.

Ethics Approval This study was performed in line with the principles of the Declaration of Helsinki.

Each project was approved by an Ethics Committee.

Contents

Acronyms

!OAIC	Online Accompanying Interactive Course
ADDIE	Analysis, Design, Development, Implementation, Evaluation
AI	Artificial Intelligence
CBL	Challenge-Based Learning
CIAO!	Corso Interattivo di Accompagnamento Online
CISIA	Consorzio Interuniversitario Sistemi Integrati per l'Accesso
CoP	Community of Practice
CRF	Centro Ricerche FIAT-Research Centre of FIAT
DAT	Differential Aptitude Test
DBL	Design-Based Learning
DigCompEdu	Digital Competence Education
EE	Engineering Education
EER	Engineering Education Researcher
EMI	English as a Medium of Instruction
FCA	FIAT Chrysler Automobiles
FSA	Fundamentals of Structural Analysis
GYM	Grow You Methodology
IDEA	Instructional Design Elementary Application
INVALSI	Istituto Nazionale per la VALutazione del Sistema educativo di Istruzione e di formazione
IoT	Internet of Things
LMS	Learning Management System
LRIC	Laboratory of Robust Identification and Control
OEF	Open Educators Factory
OEP	Open Educational Practices
OFA	Obligatory Additional Training
PBL	Problem-Based Learning
PoliTo	Politecnico di Torino
POT	Provide Opportunity in Teaching
PSVT:R	Revised Purdue Spatial Visualization Tests
RCT	Randomized Control Trial

REAL	Remote Experimentation for Automation Learning
REEN	Research in Engineering Education Network
RID	Rapid Instructional Design
SAM	Successive Approximation Model
SAperI	Spatial Ability per l'Ingegneria
SBCS	Santa Barbara Cross Section mental cutting test
SD	Standard Deviation
SMaILE	Simple Methods for Artificial Intelligence Learning and Education
SSD	Settori Scientifico Disciplinari
STEM	Science, Technology, Engineering, Mathematics
STEMM	Science, Technology, Engineering, Mathematics, Medicine
TEACH	Teaching Engineering Avant-garde Challenge Host
TIL	Test in Laib
TTPU	Turin Polytechnic University in Tashkent

Chapter 1
Introduction

In learning, you will teach, and in teaching, you will learn.

— Phil Collins

Abstract This chapter explores the intricate landscape of engineering education (EE) in Italy, with a focus on training methodologies, educational contexts, and the role of key institutions like Politecnico di Torino (PoliTo). EE is a dynamic, interdisciplinary field that integrates pedagogical strategies with engineering practices to develop skilled professionals. The Italian education system provides a robust framework, from early childhood through higher education, emphasizing STEM disciplines critical for economic and technological advancement. Comprehensive assessments such as PISA and INVALSI reveal significant trends and challenges in Mathematics and science proficiency, highlighting gender and regional disparities that require targeted interventions. PoliTo stands out as a premier technical university, offering a structured educational path and innovative approaches to engineering training. The institution's commitment to quality education is reflected in its diverse programs, rigorous admission processes, and a credit-based system that aligns with the Bologna Process. The research underscores the need for focused EE research in Italy to bridge educational gaps, enhance teaching practices, and foster a proficient engineering workforce. By addressing these areas, the chapter sets the stage for a deeper exploration of EE's impact on students and educators, aiming to elevate the standards of EE in Italy and beyond.

1.1 Context

In Italy, approximately 150,000 students graduate with a master's degree every year, and 20% of them are engineers. However, the definition of an engineer can vary widely, prompting us to ask how people are trained to become engineers, or more precisely, how engineers are trained. This fundamental query unveils a complex and multifaceted landscape that necessitates a deeper understanding.

M. G. Ballatore, *Empower Education to Foster Engineering*,
PoliTO Springer Series, https://doi.org/10.1007/978-3-031-75895-9_1

Engineering, as a discipline, is marked by remarkable diversity in its practice and applications. Consequently, the training of engineers must be adaptable, rigorous, and responsive to the evolving needs of society and industry. It is within this intricate framework that the field of engineering education (EE) emerges as a vital catalyst for change and improvement.

EE is a dynamic and multidisciplinary research domain dedicated to enhancing the quality of teaching and learning strategies in engineering. Its scope transcends traditional educational boundaries, encompassing all levels of instruction, from kindergarten to higher education. This field is characterized by a profound commitment to three core principles:

Diversity Recognizing the importance of a diverse engineering workforce, EE endeavors to broaden participation in engineering programs, ensuring that individuals from all backgrounds have the opportunity to pursue engineering careers.

Competency development EE is devoted to nurturing the competencies and skills required by modern engineers, preparing them to tackle complex challenges and innovate effectively in their respective fields.

Sustainability Sustainability, both in the educational process itself and in its alignment with societal and industrial needs, is a central concern. EE seeks to create sustainable educational practices that resonate with the evolving demands of the world.

Politecnico di Torino (PoliTo), a distinguished Italian technical university, has played a pivotal role in shaping the landscape of EE in Italy. With an annual intake of approximately 5,000 engineering bachelor's students and around 8,000 bachelor and master graduates, PoliTo stands as a beacon of EE in the country. Notably, since the academic year 2010/11, the university adopted a pioneering approach by offering a common first-year curriculum for all incoming students, regardless of their chosen major. This innovative approach allowed students to build a strong foundational knowledge base before specializing in their chosen fields.

The commitment to student-centered education at PoliTo was further underscored from 2012 onwards. These years were dedicated to enhancing teaching quality and implementing actions aimed at supporting students in their educational journey.

In 2018, a new chapter unfolded at PoliTo with the birth of the TEACH research group—Teaching Engineering Avant-garde Challenge Host. It was during this period that the idea of exploring the role and impact of engineering education researchers (EERs) in the Italian context emerged organically.

The book that follows chronicles the narrative that unfolded after these pivotal moments with critical reflections. It invites the reader to delve into the rich tapestry of EE research within the Italian context, providing insights, discoveries, and a vision for the future.

1.2 Research Objectives

In the realm of EE, the fusion of pedagogy and engineering creates a dynamic field where theory and practice converge. This unique blend of disciplines is instrumental in cultivating the essential skills required to navigate the complexities of modern engineering. It goes beyond mere academia; it is a transformative journey that molds novices into proficient engineers with a profound understanding of intricate challenges.

EE research stands as a dynamic and evolving field, harmonizing theoretical insights with practical applications. It plays a pivotal role in fostering problem-solvers who can tackle multifaceted issues, including those of ethical and societal significance. Interdisciplinarity is at the heart of EE research, nurturing awareness, ethical values, and effective communication skills, all while recognizing engineers as integral societal stewards.

In the context of Italy, there exists a noticeable void in dedicated EE research. While STEM education is gaining traction, EE often remains overshadowed. This gap is particularly striking given Italy's rich heritage in engineering, which underscores the need for dedicated research efforts. The absence of such initiatives can potentially impact the overall development of STEM education in the country. This stands in contrast to other countries such as the United States where EE research departments have been firmly established for several years now. This disparity emphasizes the necessity for Italy to further cultivate its EE research landscape to keep pace with international advancements in the field.

Filling this void is of paramount importance. It not only elevates STEM education but also ensures that engineering, a cornerstone of progress, receives the attention it rightfully deserves. Globally, there is a growing body of research contributions in EE, but practical insights specific to Italy's educators, students, and engineers are conspicuously lacking.

These pages endeavor to address this pressing issue by contributing to the strengthening of EE research in Italy. Its mission is to shape education and engineering practices, equipping educators with evidence-based methodologies, enriching student experiences, and ultimately cultivating a proficient engineering workforce capable of tackling the complex challenges of the modern world.

In summary, this research serves as a bridge to close Italy's educational gap, thereby enhancing EE. It transforms both educational and engineering practices, benefiting not only students but also society at large.

1.2.1 Main Research Question and Objectives

In the dynamic field of EE, where pedagogy and engineering intertwine, the central question guiding this research is:

What role does an engineering education researcher (EER) play within an esteemed Italian technical university like PoliTo, and how does their work positively impact both students' understanding and lecturers' roles?

This inquiry revolves around two primary objectives. Firstly, it aims to delve into the multifaceted role of an EER within the unique academic context of PoliTo. This involves a comprehensive exploration of the various dimensions and responsibilities that define the EER's role in this esteemed institution.

Secondly, this research strives to provide a compelling narrative of the tangible benefits that an EER can bring to the educational ecosystem at PoliTo. It endeavors to illustrate how the work of an EER can enhance students' comprehension of engineering concepts and simultaneously contribute to the refinement of lecturers' teaching practices.

Beyond the immediate context of PoliTo, this study holds profound implications for both the field of EE and the broader educational landscape in Italy. It unfolds in two significant dimensions.

Firstly, within the realm of educational enhancement, the research aims to contribute to the continuous improvement of EE. It draws from existing literature while adapting it to the specific Italian context. By doing so, it seeks to elevate the overall quality of the learning experiences offered to students, thus nurturing a more proficient cadre of engineers for the future.

Secondly, it emphasizes the vital role that research plays in shaping pedagogical practices. By spotlighting the pivotal role of an EER, the research underscores the intrinsic value of rigorous inquiry in the educational domain (Boyer's model of scholarship [8]). This, in turn, supports lecturers in their ongoing professional development, fostering a culture of continuous improvement in teaching.

As for the anticipated outcomes, the study will furnish empirical data that demonstrates how the work of an EER can resonate throughout the educational landscape, not merely as an abstract concept but as a tangible force for change.

Furthermore, this research aims to inform educational practices by showcasing effective strategies and interventions. By doing so, it seeks to benefit both students, who will receive a more enriching educational experience, and lecturers, who will be equipped with valuable insights to enhance their teaching methods.

In conclusion, this investigation serves as an illuminating exploration into the invaluable contributions of an EER within the Italian technical university context. It aspires to set a precedent and serve as a guiding framework for future endeavors aimed at the enhancement of EE, not only within the confines of PoliTo but as a beacon of inspiration for the entire educational landscape of Italy.

1.2.2 Methodological Framework

The methodology employed in this research is designed to systematically gather and analyze data from a diverse range of EE research projects and initiatives within the academic context. This mixed-method approach encompasses both quantitative and qualitative data, providing a holistic and inclusive representation of EE research activities at PoliTo. The aim is to offer a thorough understanding of the multifaceted impacts of EE research within the university setting, including innovative teaching methods and the development of educational technology. This approach is illustrated through the four types of scholarship outlined by Boyer [8]:

- The *scholarship of discovery*, which encompasses original research that contributes to the advancement of knowledge.
- The *scholarship of integration*, which entails the synthesis of information across different disciplines, various topics within a discipline, or over extended periods.
- The *scholarship of application* (also known as the scholarship of engagement), which extends beyond the typical service duties of faculty members to those within or outside the university, involving the rigorous application of disciplinary expertise, with outcomes that can be shared with and evaluated by peers.
- The *scholarship of teaching and learning*, which involves the systematic study of teaching and learning processes. This differs from scholarly teaching as it requires the work to be made public, available for peer review and critique according to accepted standards, and reproducible and extensible by other scholars.

The foundation of this research lies in the systematic collection of data from various projects and initiatives involving the EER at PoliTo. These projects serve as practical manifestations of the possible actions of EER, covering a wide spectrum of activities from novel teaching methods to advancements in educational technology. This comprehensive approach ensures that the research accounts for the diversity and richness of EE research within the university.

Qualitative data analysis techniques are central to this methodology. A rigorous examination of both quantitative and qualitative data from each subproject is undertaken to derive meaningful and contextually rich conclusions regarding the impact of EER activities. Thematic analysis and content analysis will be applied to uncover recurring patterns, emergent themes, and nuanced narratives from the perspectives of both students and lecturers.

The analysis goes beyond surface-level observations, delving into the underlying factors, perspectives, and experiences that shape the impact of EE research on education and pedagogy. By employing thematic and content analysis, the research aims to provide a nuanced understanding of the intricate dynamics within EE research initiatives.

This methodological framework not only assesses the current landscape of EE research at PoliTo but also offers valuable insights into potential future enhancements supported by EE research. It provides a solid foundation for deriving contextually relevant and practically applicable conclusions from the diverse array of projects under consideration.

In addition to the overarching mixed investigation approach, each project within the study follows an independent methodology tailored to its specific objectives. Projects like Matabì, SMaILE, SAperI, and DaybyDay adhere to the randomized control trial (RCT) method, ensuring a robust and controlled evaluation of their

Table 1.1 Methodology adopted for each project

Project	Methodology in short
Matabì	RCT
SMaILE app	RCT + focus group
SMaILE teachers' course	Qualitative survey
SAperI	RCT
CIAO!	Statistical analysis
TIL	Descriptive and statistical analysis
Progetto per Giovani Talenti (Honor School)	Descriptive and statistical analysis
We are HERe	Descriptive analysis, qualitative survey, and focus group
TEACH-POT	Descriptive analysis and qualitative survey
OEP	Semi-epistemological interviews
DayByDay	RCT
Math Games without frontiers	Descriptive analysis and qualitative survey
STEMM Games	Descriptive analysis and qualitative survey
IDEA ThinkLab	Descriptive analysis and qualitative survey
GYM Call	Descriptive analysis and qualitative survey
REAL Remote Lab	Statistical analysis + students' survey
PBL FAS	Statistical analysis + students' survey
Online Solutions	Descriptive analysis
W-STEM	Mixed approach
WeReLaTe	Mixed approach
Med-Mobility	Mixed approach

impact. Other projects, such as the open educational practice (OEP) initiative, utilize semi-epistemological interviews to gather in-depth qualitative insights. See Table 1.1 for the detailed list of methodologies.

This dual-layered methodology, consisting of a general mixed investigation approach and project-specific methodologies, enhances the depth and breadth of the research, allowing for a nuanced exploration of EER within PoliTo.

The implementation of this research methodology involves a carefully planned timeline to manage the intricacies of conducting multiple projects concurrently. A graphical representation (Fig. 1.1) of this timeline is provided, offering readers a visual guide to the temporal evolution of each subproject. This timeline serves as a road map, illustrating the parallel development and interconnected nature of various research run by the EER.

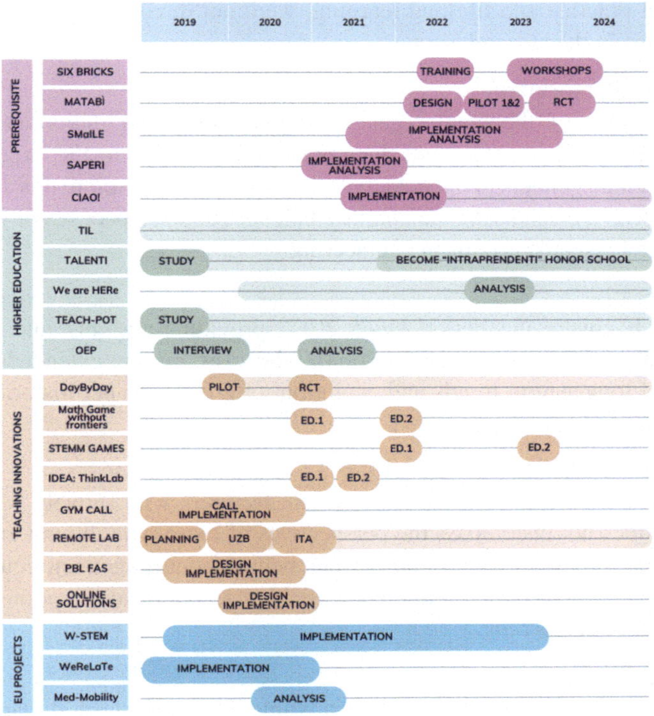

Fig. 1.1 Projects' timeline

1.3 Limits

In this section, we acknowledge and discuss the limitations encountered during the course of this research. It is crucial to recognize these limitations to provide a balanced and transparent view of the study's scope and constraints.

Sample Size and Contextual Specificity
One significant limitation of this research is the sample size. While the interactions with 47 lecturers from 10 departments provide valuable insights, the sample might not fully represent the diversity of all engineering disciplines. Additionally, the findings are specific to the context of PoliTo and may not be directly transferable to other technical universities or international settings.

Furthermore, it is important to note that these pages primarily consider projects in which the EER was directly involved. Other projects related to EE, which were autonomously implemented by faculty members, are not comprehensively examined.

Additionally, while here we address several aspects of EE research, it does not cover all relevant fields comprehensively. Some areas, such as ethics in EE or professional practice and lifelong learning, were not directly explored, indicating potential gaps in the research scope.

Gender Gap Perspective
The examination of the gender gap in engineering is an important aspect of this study. However, the research does not delve deeply into the experiences of underrepresented genders in the field. Future research could explore this aspect more comprehensively, considering the unique challenges and perspectives of women and other underrepresented groups in engineering.

Timeframe
The research covers the period since 2017, providing insights into the recent development of EE research within PoliTo. However, a more extended timeframe could reveal longer-term trends and evolution, offering a deeper historical context for the changes observed.

Interactions with EER
While the research explores interactions with lecturers who actively worked with an EER, it does not encompass the perspectives of those who did not engage with EER initiatives. Understanding the reasons behind nonengagement could provide additional insights, shedding light on potential barriers or misconceptions regarding the role of EER in teaching.

Generalizability
The findings of this study are valuable within the context of Italian technical universities, but their generalizability to other educational systems may be limited. The specific characteristics of the Italian higher education system and its challenges may not align with those of other countries. Future research could explore how EER findings and recommendations apply in diverse international settings.

Technology Integration Evaluation
The research highlights the importance of technology in education but does not extensively evaluate the effectiveness of specific technological tools or interventions. Further studies could delve deeper into the impact of technology on student learning outcomes, considering factors like the accessibility and usability of technology in different educational contexts.

Cross-Disciplinary Variations
While the study encompasses interactions with lecturers from various engineering disciplines, it does not explore potential variations in educational challenges or opportunities across different fields within engineering. Future research could investigate whether certain engineering disciplines require specialized pedagogical approaches.

Researcher Bias
Despite efforts to maintain objectivity, the researcher's perspective and involvement as an EER within PoliTo may introduce a degree of bias in the interpretation of

findings. Future research could involve multiple researchers to provide a more balanced perspective and reduce potential bias in data interpretation.

Acknowledging these limitations ensures a transparent presentation of the research's scope and constraints while providing opportunities for future studies to address these areas in more depth.

1.4 Exploring the Landscape of Engineering Education Research

Before delving into the main section, it is essential to unravel the intricate tapestry of EE and unveil its essence, significance, and multifaceted dimensions.

1.4.1 Definition

At the confluence of pedagogy and engineering, the intellectual terrain of EE research unfolds: a dynamic realm where the amalgamation of educational theory and pragmatic engineering practice imparts transformative pedagogical endeavors. The overarching goal of EE research is the systematic cultivation of knowledge, skills, and attitudes necessary for individuals to navigate the complexities of contemporary engineering landscapes. This is not merely an academic pursuit; it is an orchestrated journey that propels learners from novices to proficient engineers, armed not only with technical skills but also with a profound understanding of the intricate challenges in engineering.

Central to this educational paradigm is the fusion of theoretical foundations and experiential learning: a symbiosis that enables learners to transcend the boundaries of formulas and equations. Beyond technical competence, EE fosters cognitive transformation, developing learners into adept problem-solvers who navigate the ethical, societal, and environmental complexities inherent in their profession. EE is an interdisciplinary endeavor that nurtures holistic development.

A compendium of scholarly works collectively affirms the dynamic evolution of EE research as an interdisciplinary, international, and emerging field that remains in the process of crystallizing its identity and scope. Bernhard's exposition in 2015 supports the view that EE research can be seen as an extension of engineering research, offering engineers a unique avenue to contribute through their expertise in design, technology, and materiality [5]. This highlights the potential for engineers to significantly impact educational inquiry, shaping EE research with their deep understanding of engineering systems and their transformative effects on society.

However, Jesiek, Newswander, and Borrego's discerning observation in 2009 underscores the lingering ambiguities within the identity of EE research [19]. A lack of precise delineation about the field's objectives and aspirations persists, warranting the concerted efforts of stakeholders to elucidate and articulate the overarching goals that animate EE research. Such clarity shall engender a more cohesive trajectory for the field's evolution, facilitating informed discourse and strategic development.

Finelli and Rasoulifar's discerning perspective, articulated in 2014, resonates with the proposition that a standardized taxonomy and organizational framework are indispensable for navigating the intricate terrain of EE research [12]. An organized lexicon and structure would catalyze a more interconnected research community, fostering symbiotic collaborations among researchers and facilitating wider accessibility to research outcomes. This, in turn, could yield manifold advantages, including heightened visibility of research initiatives and a more robust network of scholarly exchange.

Building upon this intellectual foundation, Borrego and Bernhard's pronouncement in 2011 reinforces the pivotal role of EE research in preparing engineers to tackle significant engineering challenges [7]. EE research is vital for developing the competencies needed to address the technological, societal, and environmental complexities of contemporary engineering.

A constellation of seminal perspectives from Fordyce [15]; Bordogna et al. [6]; Fomunyam [14]; and Viegas et al. [27] highlight the diverse aspects of EE. While Fordyce focuses on cognitive development, Bordogna, Fromm, and Ernst advocate for integrating knowledge and human potential. Fomunyam discusses the transformative power of innovative technologies in the Fourth Industrial Revolution. Viegas, Marques, and Alves emphasize the need for EE to align with the dynamic demands of the engineering business realm, advocating for an adaptable pedagogical landscape.

In summary, EE research, as a burgeoning and globally linked domain, stands as an instrumental vanguard of the broader EE landscape. EE research, characterized by its interdisciplinary nature and resonant international impact, serves as a crucible for the fusion of educational research methods with engineering pedagogy. By acknowledging the distinctive contributions engineers offer to this evolving field, addressing the ongoing ambiguity surrounding its identity, fostering organized communication among scholars, and embracing the inexorable interplay of education, technology, and societal evolution, EE research charts a promising trajectory towards enriching the educational milieu of engineering disciplines.

1.4.2 Fields

The field of EE research encompasses a tapestry of interconnected domains that collectively converge to forge a comprehensive and nuanced understanding of the educational dynamics intrinsic to engineering disciplines. The taxonomy developed by Finelli, Borrego, and Rasoulifar in 2015, and further refined until September 2021, unfurls a plethora of fields, each characterized by distinct dimensions, foci, and research avenues [13]. This taxonomy highlights the diversity and expansiveness of EE research.

The taxonomy organizes EE research into various fields, revealing the multi-faceted efforts that researchers and educators undertake to advance engineering pedagogy. From assessment methodologies to instructional design, from diversity considerations to outcomes assessment, each field forms part of the intricate fabric of EE research. This framework serves as a guide for researchers, educators, and practitioners to navigate the expansive landscape of EE research.

Assessment This branch focuses on methodologies for organizational assessment, student performance evaluation, and program efficacy measurement. It involves traditional grading systems and innovative tools like concept inventories, promoting continuous improvement in educational quality.

Design This domain covers design practice, projects, and thinking. It involves ideation, modeling, problem definition, and transforming concepts into reality, blending creativity with methodical processes to craft innovative educational experiences. In design, we synthesize imagination with pragmatism to craft innovative educational experiences. We design the path that guides learners on their educational journey.

Diversity Addressing gender, race, underrepresentation, and workplace diversity, this field examines bias, equity, and inclusivity within EE, ensuring that educational environments promote equal opportunity.

Educational level This field addresses pedagogical strategies tailored to different educational stages, from foundational learning to advanced studies, adapting approaches to meet the evolving needs of learners.

Educational setting Focusing on classrooms, laboratories, and virtual environments, this domain optimizes learning spaces to enhance knowledge transmission and skill acquisition, making learning an immersive experience. Through optimizing educational settings, we create immersive spaces where learning is not just a process, but an experience. We curate the ambiance that harmonizes learning and inspiration.

Educational technology Exploring digital tools, online platforms, and gamified learning experiences, this field integrates technology into education to amplify its reach and impact.

Instruction This comprehensive field covers conceptual learning, faculty roles, instructional methods, and professional development, focusing on effective strategies for knowledge transmission and skill cultivation.

Outcomes Examining communication skills, critical thinking, creativity, and ethical competence, this domain assesses the ultimate goals of education and the professional trajectories of EE graduates.

Professional practice This field bridges education and industry, exploring internships, cooperative education, and real-world engineering pursuits, aligning educational outcomes with industry demands.

Recruitment and retention Focusing on student progression, this domain examines strategies for attracting and retaining students in EE programs, from academic support to recruitment initiatives.

Related fields This domain explores the intersections between EE and related disciplines, fostering cross-disciplinary understanding and innovation to enrich the educational landscape.

Research approaches Providing the methodological tools for EE research, this field includes data collection methods, research ethics, and theoretical frameworks essential for investigating EE complexities.

Theoretical frameworks Like philosophers of pedagogy, this domain delves into the philosophical foundations of education, from cognitive theories to critical perspectives, underpinning the empirical exploration of EE.

Teams Focusing on teamwork dynamics and multidisciplinary interactions, this field nurtures the collaborative competencies essential for engineering practice.

This taxonomy transcends mere categorization; it orchestrates a comprehensive exploration of EE research. Each field contributes to a collective understanding, serving as an invaluable guide for scholars, educators, and practitioners navigating EE's complexities and driving pedagogical innovation.

Transitioning into the "Context" subsection, we must explore scholarly contributions that enrich our understanding of EE research. These contributions provide the foundation upon which this taxonomy thrives, illustrating the interplay between theory and practice, innovation and tradition, and collaborative endeavors that illuminate the path toward educational excellence.

1.4.3 Context

The significance of the taxonomy of fields within EE research is deeply rooted in a multifaceted context informed by the insights of scholars navigating the intricate landscapes of education, engineering, and research methodology. This convergence of perspectives highlights the dynamic and interdisciplinary nature of EE research, as well as the rigorous methodologies and diverse knowledge artifacts that underpin its advancement.

Fortenberry [16] underscores the need for focused methodologies akin to those in the biomedical community, emphasizing tailored research methods essential for EE. This precision mirrors the complexity of EE, where optimizing pedagogy requires the same methodological rigor seen in fields like medicine.

The importance of community in the development of interdisciplinary scholars within EE research is emphasized by Allendoerfer et al. [1]. This aligns with the taxonomy's acknowledgment of interdisciplinary synergy and collaboration, illustrating how shared insights and collective exploration enrich scholars' trajectories.

Streveler and Smith [26] focus on fundamental questions about engineering learning, highlighting the transformative potential of EE research. Their shift from curriculum development to probing the core of EE aligns with the taxonomy's emphasis on exploring pedagogical paradigms and strategies to enhance learning outcomes.

The exploration of knowledge artifacts within EE research by Johri et al. [20] echoes the taxonomy's commitment to diverse perspectives. Their work on knowledge circulation, maintenance, and creation underscores the interdisciplinary intersections within the taxonomy's fields. Johri's observations on open-access online publications reflect EE research's capacity to foster accessible and collaborative knowledge dissemination.

These scholarly voices harmonize to create a contextual symphony for EE research. For instance, Madhavan et al. [22] use data mining techniques to uncover the intricate network topology within EE research, portraying it as a vibrant community of practice. Similarly, Nawaz et al. [24] define EE research through a keyword-based scheme, illustrating how scholarly exploration shapes the field' s scope. Douglas and Purzer [10] highlight the pivotal role of assessment strategies, resonating with the taxonomy's emphasis on assessment as a core thread within the educational framework.

The global dimension of EE research's growth is emphasized by Lohmann [21], who underscores its role in fostering global engineering excellence. This aligns with the taxonomy's recognition of educational settings and contexts, where international perspectives enrich the symphony of pedagogical innovation.

Collectively, these scholarly voices amplify the broader context of EE research: a dynamic, interdisciplinary realm where rigorous methodologies, collaborative ethos, and diverse knowledge artifacts converge to elevate EE. The taxonomy of fields harmonizes with these insights, showing that EE research is not a collection of isolated domains but a vibrant and evolving symphony shaping the future of education and engineering practice.

The Research in Engineering Education Network (REEN) exemplifies this collaborative ethos. REEN is an independent, international, and inclusive community of scholars dedicated to advancing the field of EE research. The community is vibrant, collaborative, and welcoming to both experts in EE research and those seeking to learn and apply sound quantitative and qualitative methodological approaches. Notably, in Italy, the TEACH group (Teaching Engineering Avand-garde Challenge Host research group) is listed as the sole representative research center in the wiki list of research centers, highlighting the need for broader recognition and inclusion of diverse research initiatives within the EE community.

1.5 Italian Context

1.5.1 Educational Overview

The Italian education system is renowned for its comprehensive structure, addressing diverse educational needs across various stages. The journey begins with early childhood education, including nursery and kindergarten programs known as "asilo nido" for children up to age 3 and "scuola dell'infanzia" for ages 3–6. These programs foster early learning and lay the foundation for primary education.

Primary education, or "scuola primaria," starts at age 6 and lasts for five years, providing a uniform curriculum nationwide. It equips young learners with fundamental knowledge and skills in subjects such as Mathematics, Italian language and literature, history, geography, science, and physical education. Teachers receive specialized training in pedagogy, child psychology, and primary education.

Following primary education is "scuola secondaria di primo grado," a three-year stage with a standardized curriculum. This level builds on primary education, introducing more advanced topics in Mathematics, sciences, literature, history, geography, and foreign languages. Teachers are trained to effectively teach specific subjects.

Upper secondary education marks a pivotal juncture where students choose between three distinct tracks, each lasting five years and culminating in a final examination. The "liceo" track focuses on a broad academic curriculum, with various sub-tracks like the Classical Lycée or Scientific Lycée, emphasizing subjects such as Mathematics. Technical schools provide practical and vocational skills, preparing students for specialized professions, while professional schools prepare students for specific trades and careers, ensuring they acquire practical competencies. Some students opt for 3- or 4-year vocational courses administered by regional authorities. The Ministry of Education sets the core curriculum for all upper secondary schools.

Italian higher education follows the Bologna Process, offering bachelor's degrees after three years and master's degrees after an additional two years. Fields such as medicine or law require longer programs, typically five to six years. The curriculum at this level is highly specialized, preparing students for their chosen professions. Professors often engage in research alongside teaching.

Compulsory education in Italy spans from age 6–16, ensuring all children enroll in educational institutions during this period. The right to receive formal training extends to age 18, demonstrating Italy's commitment to educational opportunities beyond the mandatory years. Notably, 86% of students complete secondary school, but only slightly over 50% continue to tertiary education, with just 34% attaining a degree [2, 11, 23].

Recent Italian educational policies focus on several key areas: expanding child-care opportunities for children aged 0–2, aiming to reach a 33% enrollment target, up from the current 24.7%; addressing school dropout rates, particularly high in southern regions, to lower the national average of 13.7%; and increasing the percentage of young people attaining university degrees to meet international benchmarks.

Italy places a strong emphasis on STEM education, recognizing its significance in economic and technological development. The curriculum comprehensively covers Mathematics, science, technology, and engineering, preparing students for global competitiveness.

Continuous assessment and evaluation are integral to the Italian education system, providing insights into students' progress and identifying areas for additional support. Teachers regularly evaluate performance through assignments, quizzes, oral examinations, and classroom participation. Comprehensive final examinations at the end of each secondary education stage determine eligibility for upper secondary schools and higher education. The Italian grading system uses a scale from 0 to 10, with 6 as the passing grade. In higher education, the scale ranges from 0 to 30, with 18 as the passing grade. Courses typically conclude with an examination, which may be written, oral, or both. Graduate programs often require a thesis or dissertation evaluated by faculty.

In fields such as medicine, engineering, and education, practical training and internships are essential components of the curriculum, with performance assessed based on practical experiences.

Standardized tests, designed and administered by the National Institute for the Evaluation of the Education and Training System (INVALSI), play a crucial role. These tests assess students' knowledge in key subjects at various educational stages, providing standardized measures of achievement and readiness for progression. INVALSI conducts assessments at critical points, such as the second and final years of primary school, the end of lower secondary school, and the end of upper secondary school. These tests help gauge educational outcomes and inform curriculum and policy decisions, aligning with national and international benchmarks. Further insights into INVALSI results in STEM fields are elaborated upon in the subsequent subsection.

1.5.2 STEM Field at a Glance

Within the Italian context, pivotal assessments like PISA and INVALSI provide valuable insights into the benefits of early STEM education. PISA, administered by the OECD, highlights that students exposed to rigorous STEM education from an early age perform notably better in scientific literacy, mathematical proficiency, and problem-solving skills [25]. These foundational skills not only enhance academic achievements but also have significant implications for future career prospects, particularly in engineering fields where a strong foundation in science and Mathematics is essential.

A comprehensive overview of the Mathematics proficiency assessments conducted in primary and secondary education, based on the INVALSI reports of 2023 [18], reveals important trends and insights.

1.5.2.1 Mathematics Assessment in Primary Education (II Grade)

The Mathematics assessment for second-grade students serves as an initial gauge of their numerical reasoning abilities. Analyzing trends from 2019 to 2023, the national average score has declined by 9.9 points, particularly in the last two years, likely reflecting the pandemic's impact on learning (Fig. 1.2). Approximately 64% of students achieve basic proficiency, a decrease of 7 points from 2022 (Fig. 1.3).

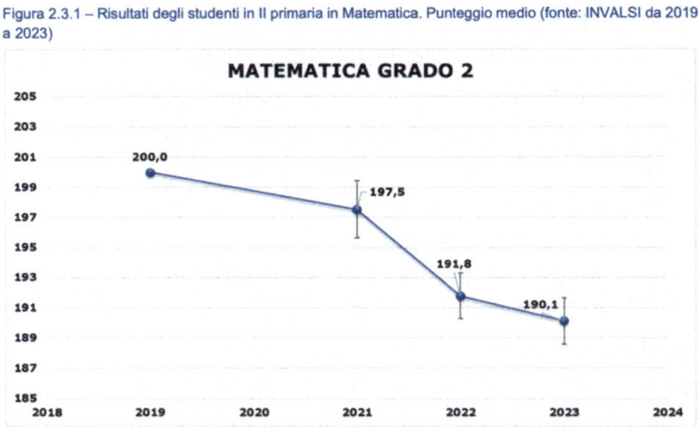

Fig. 1.2 Average score for the results for students in II primary education from 2019 to 2023 (Retrieved from [18])

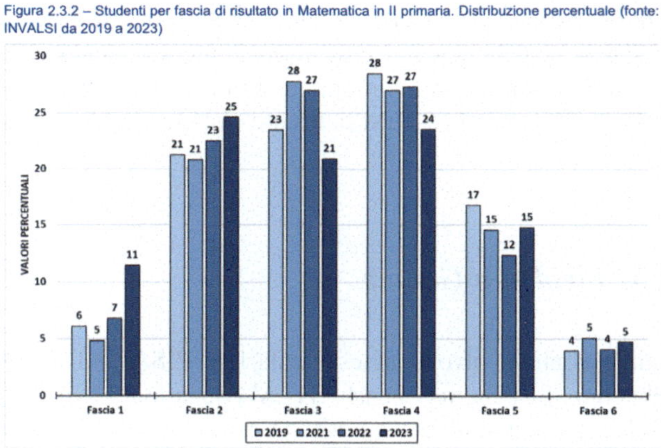

Fig. 1.3 Percentage distribution grouped by level ("fascia") for students in II primary education (Retrieved from [18])

Gender disparities are apparent, with female students scoring 6.1% points lower than males. This early gender gap warrants scrutiny and emphasizes the need for appropriate measures to address it right from the outset of students' educational journey.

1.5.2.2 Mathematics Assessment in Primary Education (V Grade)

The fifth-grade Mathematics assessment also shows a consistent decline in scores, with the national average dropping by 9.8 points from 2019 to 2023 (Fig. 1.4). Approximately 63% of students achieve basic proficiency, a 3-point decrease from 2022 (Fig. 1.5). Gender disparities continue, with female students scoring

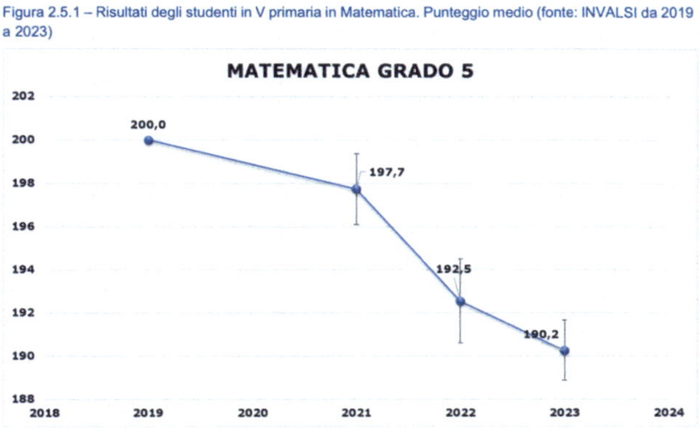

Fig. 1.4 Average score for the results for students in V primary education from 2019 to 2023 (Retrieved from [18])

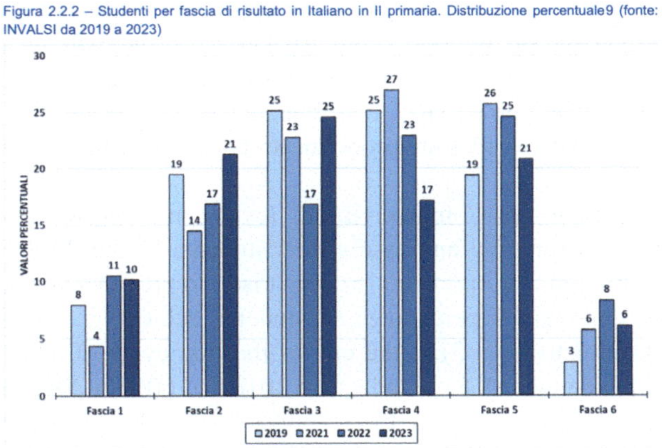

Fig. 1.5 Percentage distribution grouped by level ("fascia") for students in V primary education (Retrieved from [18])

4.6% points lower than males. Socioeconomic background significantly influences performance, with students from advantaged backgrounds scoring higher. When considering all other factors, there remains a slight difference in outcomes between different regions, with the exception of the southern and island regions where the disparity becomes significantly more pronounced (−5.2% points).

1.5.2.3 Mathematics Assessment in Lower Secondary Education (VIII Grade)

In the third grade of lower secondary education, the INVALSI Mathematics assessment measures competencies in numbers, space and figures, relationships and

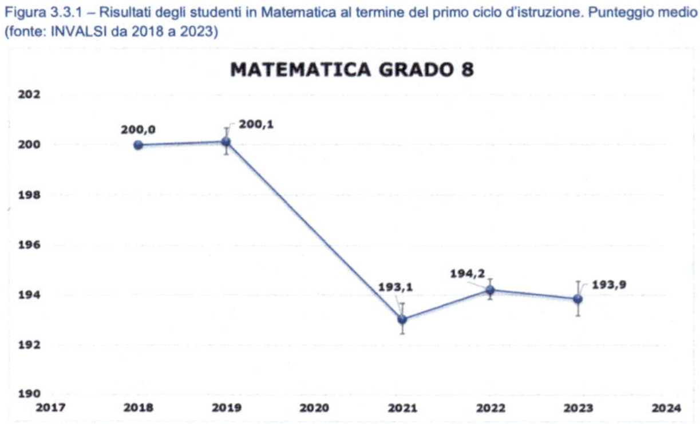

Fig. 1.6 Average score for the results for students in III lower secondary education from 2019 to 2023 (Retrieved from [18])

functions, and data and predictions. From 2018 to 2023, the national average score declined by 6.1 points (Fig. 1.6), with 55.8% of students achieving at least basic proficiency. The gender gap persists, with female students scoring 7.3 points lower than males. Regional disparities are pronounced, particularly between Northern and Southern Italy.

With the introduction of Computer-Based Testing in 2018, it becomes possible to compare results over time and assess how students in the third year of lower secondary education in 2023 perform in comparison to previous years.

Figure 1.7 monitors the percentage of students achieving at least acceptable results (from level 3 to level 5) at the end of the first cycle of education. Before the COVID-19 pandemic, 61% of students in 2018 and 60% in 2019 had reached at least the acceptability threshold. In all subsequent years, there is a decline, with no clear change in trend yet evident: 56%, a decrease of 4% points before and after the emergency.

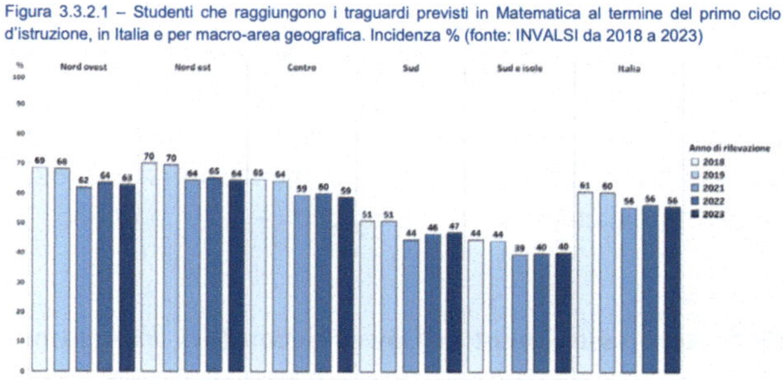

Fig. 1.7 Students who reach the expected goals in Mathematics at the end of the first cycle of education, in Italy and by geographical macro-area. Incidence %

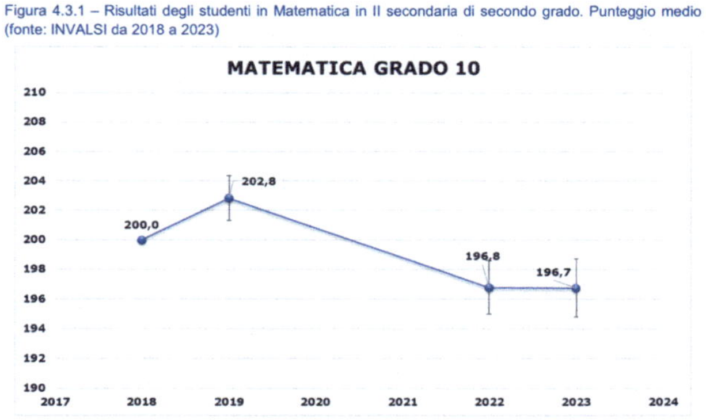

Fig. 1.8 Average score for the results for students in II upper secondary education from 2019 to 2023 (Retrieved from [18])

1.5.2.4 Mathematics Assessment in Upper Secondary Education (X Grade)

The Mathematics assessment for tenth-grade students highlights a 6.1-point decline in the national average score between 2019 and 2023 (Fig. 1.8). Only 55% of students achieve basic proficiency, with gender and socioeconomic disparities still evident. Students from scientific high schools outperform their peers by 19.7 points, while those in vocational schools lag significantly behind.

1.5.2.5 Mathematics Assessment in Upper Secondary Education (XIII Grade)

In the final year of upper secondary education, the Mathematics assessment shows a decline of 8.9 points in the national average score from 2019 to 2023 (Fig. 1.9). Half of the students achieve basic proficiency. Regional disparities are significant, with Northern Italy outperforming the south by 25.1 points, equivalent to approximately two years of learning.

Various factors influence student performance in these assessments, including gender, socioeconomic background, and regional differences. Female students consistently score lower than males, and students from advantaged backgrounds outperform their peers. Regional disparities highlight significant gaps, with Northern Italy consistently outperforming the south.

Several studies based on INVALSI data analyze the gender differences in math performance in Italy. For instance, Giberti et al. [17] suggest that these differences might be due to varying misconceptions held by males and females.

Additionally, research by Cipollone, Montanaro, and Sestito [9] indicates that value-added measures can be employed to assess the effectiveness of Italian high schools. Furthermore, their work highlights that schools in southern regions of Italy

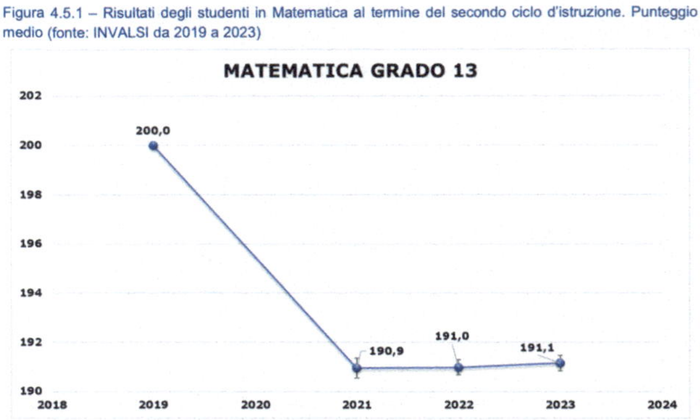

Fig. 1.9 Average score for the results for students in V upper secondary education from 2019 to 2023 (Retrieved from [18])

tend to exhibit lower starting levels of competencies and lower value-added in math and science.

Moreover, there appears to be a correlation between the performance of Italian schools in the Bebras Challenge on Informatics and Computational Thinking and their performance in INVALSI assessments, as suggested by Bellettini et al. [4].

1.5.2.6 Science Assessment in Italy

Science education in Italy is meticulously evaluated through internationally recognized assessments, specifically the Trends in International Mathematics and Science Study (TIMSS) and the PISA. These assessments provide valuable insights into what students know and their ability to apply that knowledge [3].

Italian students' performance in science lags behind many European and global counterparts. In TIMSS, fourth-grade students score an average of 516 points, while eighth-grade students score 499 points. In PISA, the average score is 468 points, below the OECD average of 489 regional disparities, and differences among educational tracks impact science performance, with academically oriented tracks performing better than technical and vocational tracks.

Additionally, the 2018 PISA round confirms and exacerbates a declining trend observed between 2012 (494) and 2015 (481), counteracting the positive trajectory initiated in 2009 (489). This decline in performance raises concerns regarding the effectiveness of current science education practices.

Gender parity in science achievement is a positive note, with no significant differences between boys and girls. However, overall student performance in international assessments, particularly in Mathematics, calls for proactive measures to bolster foundational skills and address regional and track-related disparities. These challenges underscore the importance of continuous improvements in STEM education to meet the demands of a modern workforce and foster scientific and technological advancements in Italy.

In conclusion, the examination of Mathematics and science education in Italy reveals a landscape marked by both achievements and challenges. While gender parity is a notable achievement in science education, the decline in Mathematics scores and regional disparities necessitate targeted interventions. Addressing these issues is crucial for ensuring equitable access to quality education and maintaining Italy's competitiveness on the global stage.

1.5.3 Politecnico di Torino: A Hub of Technical Education

Politecnico di Torino (PoliTo), founded in 1859, is a cornerstone of technical education in Italy, inspired by the prestigious French Ecole Polytechnique model. Specializing in engineering and architecture, PoliTo follows the Bologna model, offering a comprehensive educational structure that includes three-year bachelor's programs (Laurea), two-year master's programs (Laurea Magistrale), and three-year doctoral paths, showcasing its commitment to research excellence.

PoliTo provides a rich educational landscape with 50 distinct programs, about 40% of which are bachelor's programs spanning various disciplines in engineering (such as industrial, ICT, civil-environmental, and management) and architecture. Detailed program information is available on the official PoliTo website (www.PoliTo.it).

Aligned with the Italian university system, PoliTo employs a credit-based system requiring 180 credits for a bachelor's degree and 120 credits for a master's degree. Courses are evaluated on a score range of 18–30, with 30 indicating excellence. To progress to the second year, students must earn a minimum of 26 credits.

Admission to the architecture bachelor's program is regulated by a national level through a specific test. For engineering programs, PoliTo introduced an admission process in the 1990s, initially based on a nonselective aptitude test developed by CISIA (Consorzio Interuniversitario Sistemi Integrati per l'Accesso). In 2008/2009, PoliTo transitioned to a selective online test known as the Test in Laib (TIL), enhancing flexibility and providing immediate feedback to applicants.

As of the academic year 2017/2018, PoliTo boasts a student population of approximately 33,000, with around 4,500 freshmen in engineering bachelor's programs. The university's international appeal has grown significantly, attracting about 5,000 students from over 100 countries, constituting approximately 15% of the total student body.

PoliTo has seen a steady increase in applications for engineering bachelor's programs, contrary to national and international trends. This is further fueled by the high employment rate of its graduates, making PoliTo a global leader in graduate employability.

To manage the growing demand, PoliTo introduced a *numerus clausus* (a maximum number of admitted students) starting in the academic year 2012/2013.

This cap, set at 5,000 since 2015/2016, is determined by considering available resources and complying with national regulatory constraints.

PoliTo faces challenges related to its academic staff, with a diminishing number of professors with permanent contracts, offset by temporary contracts. This situation is influenced by factors such as the spending review and limited turnover of retired teaching staff.

The first year for engineering bachelor's students at PoliTo is foundational; it is the same for all the freshmen, covering essential science disciplines regardless of the chosen major. Subjects include Chemistry, Computer Science, Mathematical Analysis, Linear Algebra and Geometry, and Physics. These courses, forming the "anno comune," provide not only fundamental knowledge but also cultivate crucial soft skills necessary for success in engineering studies.

1.6 Book Structure

Now that we have established the context, outlined the primary research objectives, and briefly described the background, it is important to provide a road map of the structure that this book will follow. This section offers a brief overview of how the subsequent chapters are organized to systematically address the research questions and objectives.

1.6.1 Chapter 2: Nurturing STEM Pathways on the Student's Perspective

Chapter 2 embarks on a journey through the various stages of STEM education. It begins with the importance of evaluating prerequisites with entrance test (TIL).

Moving into higher education, this chapter explores programs like CIAO! and the Talenti Project, which aim to create inclusive and stimulating learning environments. It also discusses innovative approaches such as reverse inclusion and the role of laboratory activities in comprehensive understanding.

We also address the gender perspectives in EE with initiatives like SAperI that tackle gender gaps using spatial ability as a lens.

1.6.2 Chapter 3: Enhancing Engineering Education from the Lecturer's Perspective

Chapter 3 shifts the focus to the perspective of educators. It explores training opportunities and professional development, highlighting the importance of reinforcing engineering prerequisites in K-12 education and faculty development for higher education lecturers. It also delves into the concept of open educational practices (OEPs) and their potential benefits.

The chapter then delves into various course approaches and methodologies, emphasizing the role of instructional system design. It showcases examples like ThinkLab, TEACH-GYM, and innovations in course delivery.

1.6.3 Chapter 4: Discussion and Conclusion

In Chap. 4, we embark on a comprehensive discussion that serves as the culmination of our research journey. This critical chapter delves deep into the key findings and insights garnered from the preparatory aspects, technological applications, and training strategies explored in Chaps. 2 and 3. It is within this section that we synthesize the wealth of knowledge acquired, providing a holistic perspective on the transformative role of technology in EE. Through rigorous analysis and reflection, we draw meaningful conclusions that contribute to the advancement of educational practices at PoliTo and beyond, shaping the future of EE.

Finally, in this concluding chapter, we provide a concise summary of the primary findings and notable contributions stemming from this comprehensive research endeavor. Alongside this, we conscientiously address the limitations encountered in our study to ensure a well-rounded perspective.

Moreover, we emphasize the pivotal insights garnered throughout this research and propose compelling avenues for future investigations within the realm of EE, specifically within the Italian educational landscape. Our goal is to guide and inspire further exploration and development in this evolving field.

References

1. Allendoerfer C, Adams R, Bell P, Fleming L, Leifer L (2007) Becoming an engineering education researcher: finding pathways toward interdisciplinarity. In: Annual American Educational Research Association (AERA) Conference, 17
2. AlmaLaurea (2020) Condizione occupazionale dei Laureati. https://www.almalaurea.it/universita/occupazione/occupazione18/volume. Accessed: Accessed: 7 Sep 2023
3. Ballatore MG, Tabacco AM (2022) Science education in Italy. In: Science Education in Countries Along the Belt & Road. Springer, Berlin, pp 419–433. https://doi.org/10.1007/978-981-16-6955-2_25
4. Bellettini C, Lonati V, Monga M, Morpurgo A, et al (2019) An analysis of the performance of Italian schools in bebras and in the national student assessment INVALSI. In: TACKLE@EC-TEL
5. Bernhard J (2015) Engineering education research as engineering research. In: International Perspectives on Engineering Education. Springer International Publishing, Berlin, pp 393–414 . https://doi.org/10.1007/978-3-319-16169-3_19

6. Bordogna J, Fromm E, Ernst EW (1993) Engineering education: innovation through integration. J Eng Educ 82(1):3–8. Wiley. https://doi.org/10.1002/j.2168-9830.1993.tb00065.x
7. Borrego M, Bernhard J (2011). The emergence of engineering education research as an internationally connected field of inquiry. J Eng Educ 100(1):14–47. Wiley. https://doi.org/10.1002/j.2168-9830.2011.tb00003.x
8. Boyer EL (1990) Scholarship reconsidered: Priorities of the professoriate. ERIC, New York City
9. Cipollone P, Montanaro P, Sestito P (2010) L'istruzione. Il Mezzogiorno e la politica economica dell'Italia, 77
10. Douglas KA, Purzer Ş (2015) Validity: meaning and relevancy in assessment for engineering education research. J Eng Educ 104(2):108–118. Wiley. https://doi.org/10.1002/jee.20070
11. EUROSTAT (2020) Product - products datasets - eurostat . https://ec.europa.eu/eurostat/web/products-datasets/-/sdg_04_10. Accessed: 7 Sep 2023
12. Finelli CJ, Rasoulifar G (2014) Using and disseminating a taxonomy for engineering education research. In: 2014 IEEE Frontiers in Education Conference (FIE) Proceedings. IEEE. https://doi.org/10.1109/FIE.2014.7043989
13. Finelli CJ, Borrego M, Rasoulifar G (2015) Development of a taxonomy of keywords for engineering education research. J Eng Edu 104(4):365–387. Wiley. https://doi.org/10.1002/jee.20101
14. Fomunyam KG (2020) The language of engineering education in the era of the fourth industrial revolution. Int J Innovat Technol Expl Eng 9(9):518–523. Blue Eyes Intelligence Engineering and Sciences Publication - BEIESP. https://doi.org/10.35940/ijitee.i7160.079920
15. Fordyce D (1986) Engineering education: a total concept? Assessm Evaluat Higher Educ 11(3):240–256. Informa UK Limited. https://doi.org/10.1080/0260293860110308
16. Fortenberry NL (2006) An extensive agenda for engineering education research. J Eng Educ 95(1):3–5. Wiley. https://doi.org/10.1002/j.2168-9830.2006.tb00872.x
17. Giberti C, Viale M, et al (2019) L'impatto del gap linguistico nelle performance degli studenti madrelingua e non madrelingua italiana: dai risultati delle prove INVALSI al lavoro in classe. INVALSI per la Ricerca, 117–140. Franco Angeli
18. INVALSI. (2023). Rapporto INVALSI 2023. INVALSI Roma
19. Jesiek BK, Newswander LK, Borrego M (2009) Engineering education research: discipline, community, or field? J Eng Educ 98(1):39–52. Wiley. https://doi.org/10.1002/j.2168-9830.2009.tb01004.x
20. Johri A, Edstrom K, Du X, Mitchell J, May D (2022) Knowledge construction in engineering education research – assessing the role of journals, books, conferences, and other products of research. In: 2022 IEEE Frontiers in Education Conference (FIE). IEEE. https://doi.org/10.1109/FIE56618.2022.9962543
21. Lohmann JR (2009) Global engineering excellence: the role of educational research and development. Revista de Ensino de Engenharia 27(3):33–44
22. Madhavan K, Xian H, Johri A, Vorvoreanu M, Jesiek BK, Wankat PC (2011) Understanding the engineering education research problem space using interactive knowledge networks. In: 2011 ASEE Annual Conference & Exposition, pp 22–1574
23. MIUR (2020) Esplora Dati. https://dati.istruzione.it/espscu/index.html?area=anagScu. Accessed: 7 Sep 2023
24. Nawaz S, Rajan P, Yu JH, Luo Y, Choi JH, Radcliffe DF, Strobel J (2011) A keyword based scheme to define engineering education research as a field and its members. In: 2011 IEEE Global Engineering Education Conference (EDUCON). IEEE. https://doi.org/10.1109/EDUCON.2011.5773137
25. OECD Indicators (2016) Education at a Glance 2016. Editions OECD, 90
26. Streveler RA, Smith KA (2006) Conducting rigorous research in engineering education. J Eng Educ 95(2):103–105. Wiley. https://doi.org/10.1002/j.2168-9830.2006.tb00882.x
27. Viegas C, Marques A, Alves G (2018) Engineering education and technological/professional learning. In: Proceedings of the Sixth International Conference on Technological Ecosystems for Enhancing Multiculturality. ACM. https://doi.org/10.1145/3284179.3284192

Chapter 2
Nurturing STEM Pathways on the Student's Perspective

Two roads diverged in a yellow wood,
And sorry I could not travel both
And be one traveler, long I stood
And looked down one as far as I could
To where it bent in the undergrowth;
Then took the other, as just as fair,
And having perhaps the better claim,
Because it was grassy and wanted wear;
Though as for that the passing there
Had worn them really about the same,
And both that morning equally lay
In leaves no step had trodden black.
Oh, I kept the first for another day!
Yet knowing how way leads on to way,
I doubted if I should ever come back.
I shall be telling this with a sigh
Somewhere ages and ages hence:
Two roads diverged in a wood, and I—
I took the one less traveled by,
And that has made all the difference.

—Robert Frost, The Road Not Taken

Abstract In STEM education, the journey of aspiring engineers is a tapestry woven with preparation, discovery, and diversity. The pivotal role of students in this narrative cannot be overstated—they are the architects of their own learning journeys, the seekers of knowledge, and the future torchbearers of innovation. As they traverse the educational landscape, their experiences shape the foundation of their engineering aspirations and contributions to society.

The chapter is structured to encompass a spectrum of insights, each contributing to a holistic understanding of the student's role in EE. It embarks on higher education, where the convergence of methodologies, projects, and inclusive initiatives culminate in a rich tapestry of learning. Amidst these endeavors, the chapter introduces the gender dimension in engineering, recognizing the significance of a balanced and inclusive representation.

M. G. Ballatore, *Empower Education to Foster Engineering*,
PoliTO Springer Series, https://doi.org/10.1007/978-3-031-75895-9_2

As the chapter unfolds, it underscores the collective impact of these stages, highlighting that EE is not just a transmission of knowledge, but a transformative journey that shapes the intellect and perspective of each student. Through this exploration, the chapter aims not only to showcase the challenges and triumphs faced by students but also to celebrate their indomitable spirit and potential to engineer a better future.

This chapter delves into the multifaceted world of engineering education (EE) from the student's perspective. It sheds light on the transformative stages that students undergo, beginning with the preparatory phase that lays the cornerstone for their future pursuits. From there, the exploration extends to the terrain of higher education, where students delve deeper into their chosen disciplines, honing their skills and fostering a spirit of innovation. An essential component of this journey is the exploration of gender perspectives, highlighting the importance of diversity in EE and its potential to reshape the landscape.

Within these pages, we invite readers to delve into the intricate narratives of students' aspirations, triumphs, and contributions in the realm of EE. Join us as we navigate the paths of preparatory studies, higher education pursuits, and inclusive perspectives, uncovering the threads that weave the fabric of excellence in EE.

2.1 Approaching Gender Gap Using Spatial Ability: SAperI

Spatial ability is a multifaceted cognitive skill that encompasses various intellectual competencies related to the representation, transformation, generation, and recall of nonlinguistic information. Historically, spatial ability has been associated with success in STEMM fields, making it a crucial factor for individuals aspiring to pursue careers in engineering, science, drafting, designing, and related disciplines [15, 17, 32, 35]. The proficiency in spatial ability is essential for tasks such as interpreting graphs, solving complex part-whole relations, and visualizing mathematical models in real-world scenarios, skills that are particularly crucial in STEMM careers [29, 42].

The existence of a strong association between spatial ability and success in STEMM disciplines has been supported by various studies. It has been shown that individuals who demonstrate high spatial ability tend to perform well in Mathematics and scientific thinking [22, 26, 38], standardized tests (SATs) [18], and occupational aptitude tests, indicating its predictive validity for STEMM career paths [24]. Longitudinal studies have further confirmed this association, following participants from high school through university and their subsequent careers, revealing a consistent link between spatial ability and science and engineering career preferences [48].

However, despite the importance of spatial ability in STEMM fields, a significant gender gap has been observed in this cognitive domain. Robust gender differences exist in spatial ability, with males generally outperforming females on certain spatial tasks, particularly those related to mental rotation [20, 25, 27, 30, 34, 39, 43, 47]. On the other hand, smaller gender differences are observed in spatial perception tasks [31, 32]. These gender disparities in spatial ability have raised important questions about their implications for the representation of women in engineering and other STEMM disciplines.

The gender gap in spatial ability has been a subject of extensive research, and multiple factors have been proposed to explain its existence. Among these are psychological factors, societal stereotypes, and environmental influences [7, 16, 23, 35–37]. For instance, girls may perceive spatial tasks as masculine and may be more intimidated by them due to prevailing gender stereotypes [37]. Additionally, the lack of support and encouragement from professors, families, and peers, as well as the absence of female role models and mentors in STEMM fields, might contribute to the persistence of the gender gap.

Addressing the gender gap in STEMM fields requires a comprehensive understanding of the role of spatial ability in career choices and success. Various initiatives have been designed to enhance girls' spatial ability and prepare them for future careers in engineering and related disciplines [45]. Such interventions aim not only to bridge the gender gap in spatial ability but also to create a supportive network that challenges stereotypes and prejudices in society, fostering an inclusive and diverse environment in STEMM fields.

In this subsection, we delve into the relationship between spatial ability and the gender gap in engineering. We explore the significance of spatial ability in STEMM disciplines, the impact of the gender gap on career choices, and the potential of interventions like Sorby course or the SAperI summer school to empower young women in pursuing engineering and other STEMM careers. By examining existing research and drawing insights from the SAperI initiative, we aim to shed light on the challenges and opportunities in addressing the gender gap through spatial ability enhancement programs in the field of engineering.

The Sorby course, developed by Dr. Sheryl Sorby and her colleagues, is a significant intervention designed to enhance spatial ability and promote diversity in STEMM disciplines, particularly in engineering and related fields [46]. The course aims to address the gender gap in spatial ability by providing targeted training and support to female students, empowering them to pursue STEMM careers and challenging the stereotypes that hinder their participation in these fields.

The Sorby course represents a proactive intervention to address the gender gap in spatial ability and promote diversity in STEMM disciplines. The course was initially introduced as a 3-credit program called "Introduction to Spatial Visualization" at Michigan Technological University. A pilot course was ten weeks long and required the participants to get engaged in hands-on construction activities, paper and pencil exercises, and computer-based tasks designed to develop their 3-D spatial skills. Then, maintaining the contents, a new semester course was designed [45] along with the accompanying multimedia software. The RCT study reveals that it is just as effective in improving spatial skills as the previous quarter system course that utilized traditional lectures.

The positive outcomes of the Sorby course extended beyond the improvement in spatial skills. Students who participated in the course performed significantly better in their subsequent engineering courses compared to their counterparts in the control group. Additionally, the course had a positive impact on retention rates in engineering. Students in the experimental group were retained in engineering at a higher rate than those in the control group.

An interesting observation from the evaluation is that the differences in retention rates for women in the experimental group were significantly higher than for women in the control group. This indicates that the Sorby course had a particularly empowering effect on female students, helping them stay engaged in engineering and increasing their representation in the field.

Following the Sorby course's positive impact on spatial abilities and engineering career outcomes, we now turn our attention to the "SAperI–Spatial Ability per l'Ingegneria" summer school [9](in English: Knowledge–Spatial Ability for Engineering), an intense summer school for 17-year-old girls. It was organized by Politecnico di Torino (PoliTo) as part of the *We are HERe* campaigns. SAperI was a targeted initiative aimed at exploring the potential of interventions in improving spatial abilities, particularly among young women, and encouraging them to pursue engineering and other STEMM disciplines. In the meantime, from the research point of view, through the school we look at the effectiveness of an intense five-day training program in improving spatial abilities among the participants.

The summer school took place during the summer of 2019 and was open to female high school students in the Piedmont and Valle d'Aosta regions of Italy (Fig. 2.1). PoliTo leveraged its "Progetto Orientamento," a well-established network of high schools, to reach out to potential participants. Out of the 167 students from 11 high schools who expressed interest, 93 female students and 74 male students participated in the initiative.

After the selection process, 37 female students completed all phases of the SAperI summer school, making them the focus of the evaluation. The course was structured to be intensive and immersive, designed to improve spatial abilities through various engaging activities and exercises.

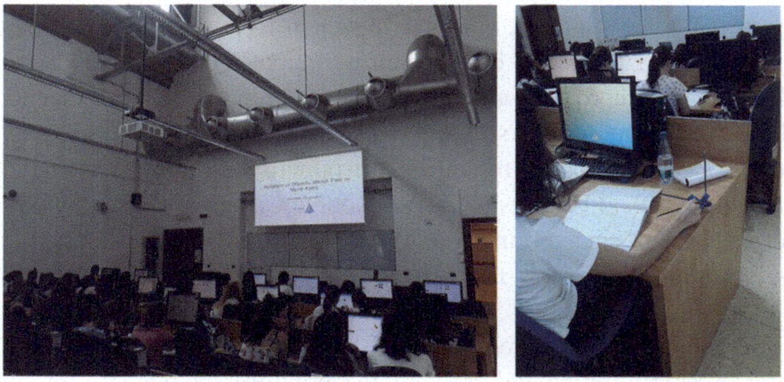

Fig. 2.1 Pictures of the SAperI summer school

The SAperI summer school consisted of a carefully planned curriculum, covering five modules based on the course developed by Sorby [46]. Each module was introduced by a moderator, and the students engaged in pairs to explore the topics through computer-based exercises. They also completed workbook exercises, and alternative solution strategies were shared at the end of each session.

In addition to the morning sessions, which focused on the Sorby course modules, the afternoons were dedicated to "play and learn" activities. The participants were divided into homogeneous groups based on preliminary test results and named after female STEM role models. The afternoon activities included:

- Trail-O: a desk-based version of Orienteering, with 5 games of increasing difficulty
- Graphical problem-solving: individual problem-solving exercises followed by a group competition
- Origami e Touch Sketch: a lesson on technological origami followed by a final competition

The evaluation of the SAperI summer school was based on three testing phases: the preliminary evaluation in May 2019 (Phase a), the evaluation at the end of the summer school in June 2019 (Phase b), and the final evaluation in September 2019 (Phase c). To assess the participants' spatial abilities, standardized testing instruments were utilized, including the Revised Purdue Spatial Visualization Tests (PSVT: R) [49], Modified Lappan Spatial Visualization Assessment (Lappan) [28], Santa Barbara Cross Section mental cutting test (SBCS) [19], and Paper Folding component of the Differential Aptitude Test (DAT)[14]. At the end of the series of tests, each student could see his/her personal score in each test. In general, males outperformed females in all four tests (Table 2.1). As the mean spatial ability score of the participants that actually took part at SAperI summer school (experimental group) is slightly higher than those who were selected, it seems that those who opted not to take the course had slightly lower spatial ability.

Table 2.2 presents the average test results for each evaluation phase (a, b, and c) in all four spatial ability tests. The data clearly indicate a substantial improvement in spatial abilities from the initial evaluation (Phase a) to the evaluation at the end of the summer school (Phase b) for all four tests. This trend of improvement persisted even in the final evaluation (Phase c), confirming the durable impact of the SAperI summer school on spatial abilities.

Table 2.1 Descriptive statistics of the pretest (Phase a) considering the different populations involved in the SAperI evaluation

	N	Revised PSVT:R	Lappan	SBCS	DAT
Males	74	M = 19.15 SD = 5.84	M = 5.73 SD = 2.46	M = 19.47 SD = 6.82	M = 7 SD = 2.55
Females	93	M = 16.23 SD = 4.78	M = 4.45 SD = 2.02	M = 17.13 SD = 7.72	M = 6.62 SD = 2.17
Selected	46	M = 16.61 SD = 5.32	M = 4.93 SD = 2.13	M = 18.15 SD = 7.13	M = 6.80 SD = 2.33
Who took part	37	M = 17.14 SD = 5.37	M = 5.05 SD = 2.21	M = 18.46 SD = 7.00	M = 7.03 SD = 2.17

Table 2.2 Descriptive statistics of the results achieved by the SAperI summer school participants

Test	Step (a)	Step (b)	Step (c)
Revised PSVT:R	Min=6	Max=27	M=17,49
	Min = 13	Max = 29	M = 21,94
	Min = 13	Max = 29	M = 22,77
	SD = 5,22	SD = 4,07	SD = 3,94
Lappan	Min = 1	Max = 9	M = 5,2
	Min = 2	Max = 10	M = 7,34
	Min = 3	Max = 10	M = 7,97
	SD = 2,10	SD = 1,85	SD = 1,98
SBCS	Min = 7	Max = 29	M = 18,77
	Min = 14	Max = 30	M = 24,20
	Min = 9	Max = 30	M = 24,69
	SD = 6,91	SD = 3,78	SD = 4,21
DAT	Min = 3	Max = 10	M = 7,28
	Min = 4	Max = 10	M = 8,51
	Min = 5	Max = 10	M = 8,54
	SD = 1,92	SD = 1,58	SD = 1,44

The average gain scores for all four tests indicate a positive improvement from Phase (a) to (b). However, it is essential to note that for the SBCS and DAT tests, a small number of participants scored lower results from Phase (a) to (b). This observation may be attributed to the intensity of the morning sessions, which could have affected their performance during the afternoon evaluation. Another explanation could be that the flat pattern module was the last one to be taught and they could have had less time to elaborate the content.

As mentioned by Sorby and Baartmans [46], considering the two-month gap between the tests, any gains in scores due to practice effects are expected to be minimal. To ascertain whether the score improvements are indeed attributed to the summer school rather than mere repetition of the tests, we can compare the ratios of the gain in points from Phase (a) to (b) (Table 2.3) and from Phase (b) to (c) (Table 2.4).

The previously specified ratio falls within the range of 0.72–1.30 for Phase (a) to (b) and 0.02–0.33 for Phase (b) to (c). Clearly, the ratio pertaining to the gain from the initial evaluation to the evaluation at the end of the summer school surpasses the latter. This observation provides compelling evidence that SAperI has significantly contributed to the improvement of participants' spatial abilities.

Table 2.3 Average gain scores from Phase (a) to Phase (b)

	Gain	Standard deviation	t-Value	p-Value	Gain (in function of)
PSVT	4,457143	3,424332	3,985078	0,000175	1,301609
LAPPAN	2,142857	1,768064	4,536763	0,000024	1,211979
SBCS	5,428571	5,658340	4,077006	0,000155	0,959393
DAT	1,228571	1,699234	2,925381	0,004721	0,723015

Table 2.4 Average gain scores from Phase (b) and Phase (c)

	Gain	Standard deviation	t-Value	p-Value	Gain (in function of)
PSVT	0,828571	2,479157	0,86493	0,390122	0,334215
LAPPAN	0,628571	1,415995	1,374526	0,173813	0,443908
SBCS	0,485714	3,08098	0,508172	0,612996	0,157649
DAT	0,028571	1,042782	0,079057	0,937221	0,027399

Moreover, the average test scores during Phase (c), along with the corresponding p-values for the data set of Phase (b) and (c), are all substantially higher than the significance level. This further validates the enduring impact of the SAperI summer school, indicating that the improvement in spatial abilities is sustained over time.

Beyond the quantitative evaluation, participants' feedback on the SAperI summer school was overwhelmingly positive. The survey conducted at the end of the program revealed that the majority of participants considered it an extremely positive or positive experience. Many of the girls stated that SAperI had influenced their university choices, motivating them to pursue STEMM degrees.

The long-term impact of the SAperI summer school was assessed through a follow-up survey in May 2020. Among the respondents, 87% were attending higher education, with 77% pursuing STEM degrees. Notably, 37% of the participants believed that SAperI had a direct influence on their university choice.

The evaluation of the SAperI summer school demonstrates its effectiveness in enhancing spatial abilities among young women and fostering interest in engineering and other STEMM disciplines. The significant improvements in spatial test scores, as well as the positive feedback from participants and their subsequent career choices, highlight the importance of targeted interventions to bridge the gender gap in STEMM fields.

The findings further reinforce the potential of spatial ability enhancement programs, such as the Sorby course-based SAperI summer school, in empowering young women and challenging gender stereotypes in engineering. By investing in such initiatives and creating inclusive learning environments, we can pave the way for a diverse and thriving future in engineering and other STEMM disciplines.

2.2 Evaluate the Competence Background of Potential Engineering Students: TIL

Preparatory engineering studies encompass a blend of formal and informal learning modalities, each contributing to the development of STEM-oriented minds. These learning approaches encompass diverse techniques and elements that support the cultivation of competencies essential for potential engineering students. Considering the heterogeneous range of background competencies requisite for success in EE, a meticulous evaluation of students' preparation becomes imperative. Understanding the diverse nature of individuals' STEM acquisition pathways enables educational institutions to implement comprehensive assessment strategies that consider both formal and informal learning contexts. Such an extensive evaluation process ensures the identification and adequate preparation of students possessing the necessary competencies to navigate the challenges inherent in an engineering curriculum. This rigorous assessment approach plays a pivotal role in fostering excellence in EE and nurturing a cohort of proficient and adept engineering professionals.

As said, assessing prospective engineering students' competencies before enrollment in a university has become crucial in ensuring a successful academic journey. This section focuses on the Test in Laib (TIL) solution and its significance in evaluating the competence background of engineering students. The primary objective of the TIL is to assess the competence background required for a successful career in a technical university. The TIL has been developed to address the need for an effective and robust evaluation process associated with merit, as well as to enhance the attractiveness and quality of incoming students while reducing the dropout rate.

The TIL is an innovative tool designed to evaluate potential engineering students' competence background at the PoliTo. It was developed in response to the introduction of the numerus clausus and the growing demand for a robust evaluation process based on merit. The TIL consists of 42 multiple-choice questions divided into four sections: Mathematics, Physics, Verbal Comprehension, and Logic. This comprehensive test provides a reliable and predictable means of assessing the competence background required for a successful career in a technical university. Additionally, the TIL incorporates a self-assessment feature that allows applicants to make informed decisions regarding their individual choices.

The TIL analysis [3] employed a data-driven approach and demonstrates the effectiveness of this assessment tool in evaluating and improving STEM learning outcomes. The TIL analysis involved integrating data from various sources, including the students' population and the total number of enrolled students. By comparing test results with students' subsequent academic careers, it is evident that the TIL is a reliable predictor of students' performance. The introduction of the numerus clausus and the implementation of the TIL test have significantly improved the competence background of students entering engineering programs. This, in turn, has increased incoming students' overall attractiveness and quality while maintaining a high employability rate.

The test was designed to evaluate aptitude and assess the effectiveness of previous preparatory studies. The scoring system awards 1 point for each correct answer while deducting 0.25 points for incorrect answers. The strong correlation between test results and indicators of students' success, such as full earned credits (Fig. 2.2a), average exam scores (Fig. 2.2b), and global career duration, emphasizes the predictive power and validity of the TIL. This correlation reinforces the TIL's predictive power and underscores its role in identifying students who are likely to succeed in their university careers.

The TIL analysis highlights the tool's effectiveness in evaluating and improving STEM learning outcomes. The results indicate that introducing a numerus clausus and a selective test, such as the TIL, have improved the competence background of

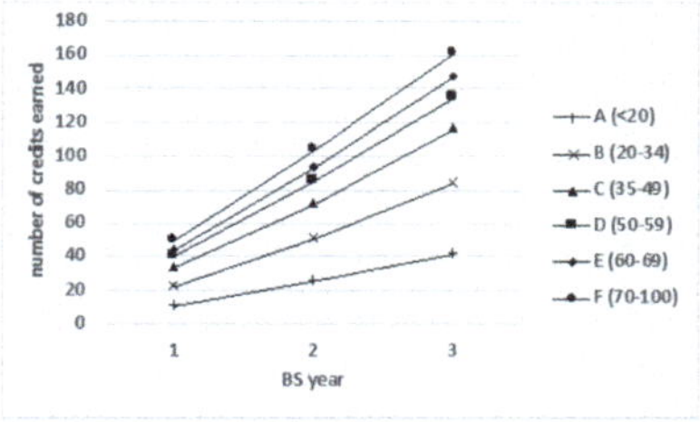

(a)

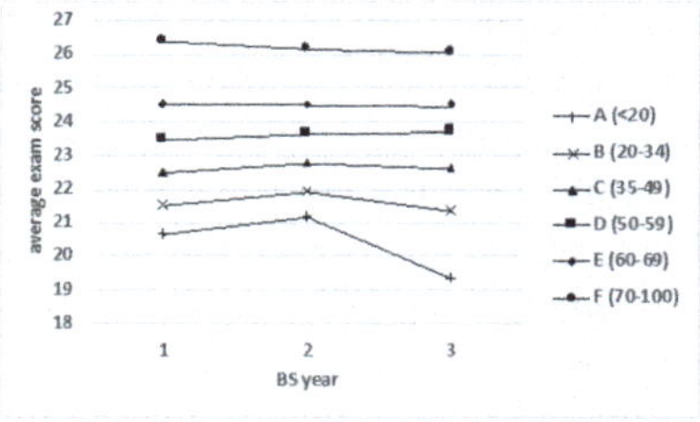

(b)

Fig. 2.2 Similar trend is observable correlating the average number of credits and the average exam scores at the end of each BS year to the TIL result. (**a**) Correlation between the total number of credits earned and the TIL score category. (**b**) Correlation between average exam scores and TIL scorecategory

incoming engineering students. Furthermore, the TIL has demonstrated its ability to predict students' university career outcomes, as evidenced by the correlation between test results and indicators of success. Establishing a minimum test score requirement has contributed to enhancing the overall quality of first-year students and reducing the dropout rate through targeted selection at the entrance stage.

In a.y. 2022/2023 the TIL composition has been slightly modified reducing the logic and reading sections and adding a new section focused on technology. This includes questions related, for example, to basic programming and spatial abilities. We are currently studying the validity of the revised TIL.

In conclusion, the TIL represents an effective solution for evaluating the competencies of potential engineering students. By assessing the competence background before enrollment, the TIL contributes to improving STEM learning outcomes, reducing the dropout rate, and enhancing the quality of incoming students. The TIL has demonstrated its reliability, robustness, and predictive capability in selecting engineering students, thereby facilitating their successful academic journey and enhancing their employability in technical fields.

2.3 Addressing the Missing Prerequisite: CIAO!

Building upon the comprehensive evaluation process highlighted by the TIL discussed in the previous subsection (Sect. 2.2), it becomes evident that assessing the competence background of prospective engineering students before university enrollment plays a pivotal role in ensuring a successful academic journey. The TIL has successfully addressed the need for a robust evaluation process based on merit, ultimately enhancing the quality and attractiveness of incoming students while reducing the dropout rate. However, competence assessment goes beyond the initial evaluation and requires ongoing efforts to bridge the knowledge gap and ensure a smooth transition into higher education. Recognizing the heterogeneous range of background competencies requisite for success in EE, providing students with effective means to fill any existing knowledge gaps is imperative. In this context, the Corso Interattivo di Accompagnamento Online (CIAO!), or, in its English version, Online Accompanying Interactive Course (OAIC!), serves as an initiative aiming to support incoming students by providing interactive and engaging materials that facilitate the acquisition of fundamental mathematical prerequisites. By addressing the knowledge gap before commencing the core curriculum, CIAO plays a crucial role in promoting a solid foundation of mathematical competencies and fostering academic success among engineering students.

The CIAO is an innovative project developed at the PoliTo in 2021 to provide interactive and personalized support to incoming students to address any deficiencies in their mathematical knowledge before commencing the core curriculum. The primary goal of CIAO is to enable students to acquire a solid foundation of mathematical prerequisites by promoting active and engaging learning.

The course is accessible through the university's Moodle platform. It offers a wide range of interactive materials, including interactive e-books with videos, interactive exercises with immediate feedback, concise theoretical texts, games, and definitions implemented using H5P (Fig. 2.3a and b). One distinctive feature of CIAO is the adoption of game-based learning principles, allowing students to accumulate experience points during their interactions with the course content (Fig. 2.4) and engage in a final challenge within a gaming environment represented by an escape room (Fig. 2.5).

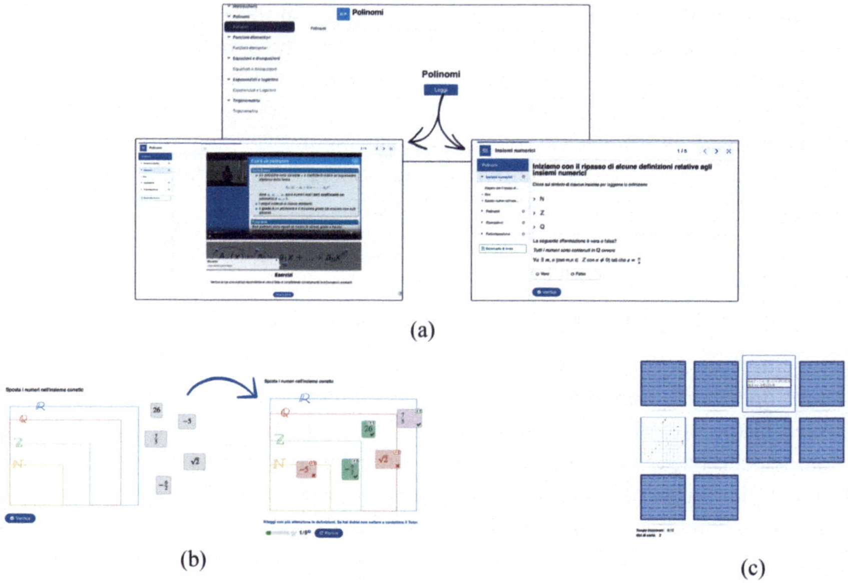

Fig. 2.3 Screenshot of CIAO! course. (**a**) A book with some examples of video and interactive materials. (**b**) Example of interactive game with immediate feedback. (**c**) Learning the domain of functions using memory

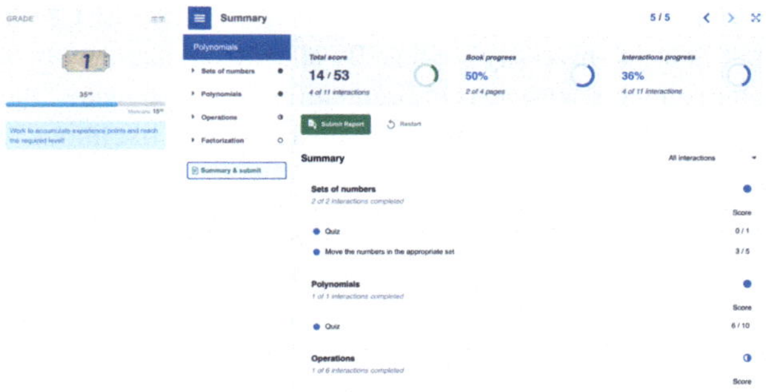

Fig. 2.4 Gamification elements in CIAO! course

Fig. 2.5 Screenshot of escape room

In September 2021, the CIAO was initially introduced solely in Italian and was not compulsory for students. However, recognizing the need to cater to an international student body, an English version of CIAO was subsequently developed and made available in 2022. Furthermore, as of September 2023, CIAO has transitioned into an Obligatory Additional Training (OFA) module, requiring all students who receive a low score in the Mathematics section of the TIL to complete the CIAO program. This mandatory implementation of CIAO aims to address and bridge the knowledge gap in mathematical competencies among students, ensuring a solid foundation before embarking on their EE journey. In general, the key objectives of CIAO include the following:

- *Remediation of mathematical deficiencies*: The course aims to provide students with tools and resources to address any gaps in their mathematical skills. CIAO seeks to reinforce the fundamental mathematical knowledge required for successful university studies through an interactive and engaging approach.
- *Promoting active learning*: By employing innovative teaching methodologies such as game-based learning, CIAO encourages students to participate in the learning process actively. The interactivity of the materials, immediate feedback, and moments of assessment within the gaming environment stimulate student interest and foster active engagement in Mathematics.
- *Enhancing student confidence and self-assurance*: Through successfully completing challenges presented by CIAO and achieving positive outcomes, students develop increased confidence in their mathematical abilities and a greater sense of self-assurance in their choice of study. This translates into greater active participation and a higher likelihood of academic success.

- *Reducing early attrition rates*: Another crucial objective of CIAO is to reduce early dropout rates in university studies. By addressing mathematical deficiencies and equipping students with a solid foundation of prerequisites, the course aims to diminish feelings of frustration and decrease the likelihood of premature discontinuation.

The CIAO! is structured into five modules: polynomials, elementary functions, equations and inequalities, exponentials and logarithms, and trigonometry. Each module is presented through an interactive book implemented using H5P technology. The interactive books incorporate various elements such as short videos, theoretical content, and diverse exercises aimed at engaging students actively in the learning process. Upon completing each exercise, students receive immediate feedback presented in a gamified visual format, where a moving star indicates their score on a progress bar, accompanied by an explanatory message.

Furthermore, CIAO! utilizes a game-based learning approach to enhance student motivation and participation. As students interact with the course content, they accumulate experience points, allowing them to progress to higher levels. Each level attained unlocks a "lottery ticket" (Fig. 2.4) allowing students to participate in unique draws. Additionally, CIAO! features an intriguing challenge in the form of an online escape room designed to promote self-evaluation and consolidate knowledge within a playful environment. The escape room scenario involves solving enigmas related to CIAO topics, with the ultimate goal of successfully navigating through different locations, such as the Aula Magna, courtyard, and main hall, and ultimately finding the correct classroom and seat (Fig. 2.5). Students are given 24 hours to complete the challenge and evaluate their understanding of the course material. As a reward for completing the escape room, prizes are drawn based on the number of accumulated lottery tickets.

While students have various means to address missing prerequisites, such as self-study, face-to-face courses, tutoring sections, MOOCs, and online resources, it has been observed that the incorporation of playful learning experiences and gamification elements significantly increases students' interest in bridging their knowledge gaps. Introducing a gamified approach within CIAO incentivizes students to actively engage with the course material, fostering a more enjoyable and practical learning experience. This approach not only promotes knowledge acquisition but also enhances students' intrinsic motivation, thereby contributing to a more comprehensive and successful preparation for their engineering studies. Considering the positive results achieved, CIAO has expanded its scope and now includes the prerequisite assessment of other subjects, such as physics, further enhancing its effectiveness in preparing students for their engineering journey.

2.4 An Inclusive and Stimulating Environment: Talenti Project

Starting university marks a significant milestone in students' academic journey, where they embark on a path of higher education and specialized knowledge acquisition. At PoliTo, guided by the indispensable assumption of the absolute centrality of the student, comprehensive support and services have been provided to students throughout their educational journey. As part of this commitment, the initial year is dedicated to establishing a solid foundation in key knowledge areas, including Mathematical Analysis, Computer Science, Chemistry, Physics, Algebra, and Geometry. The university offers various teaching support initiatives to facilitate the transition from high school to university, such as realignment courses for students with specific gaps in Mathematics and Physics (like the CIAO program mentioned in the previous Sect. 2.3), recovery courses for those who did not pass the initial Mathematics exam, recorded lessons available online, and tutoring assistance.

While these initiatives cater to the needs of a significant portion of the student population, it is essential to further focus on the quality of students' preparation and take specific actions to cater to a segment of the student population exhibiting exceptional skills. A technical university must explore and implement innovative approaches to strengthen the learning processes, particularly during the bachelor's years, in light of evolving didactic potentials and social requirements. In this context, the board of PoliTo has posed several fundamental questions: How can a technical university effectively support talented students? Can this process be inclusive without completely segregating them from their peers? Additionally, what is the impact of these actions on the overall student population, including students with difficulties and those following a standard career trajectory?

To address these inquiries, we delve into the examination of the "Research among Quality-Young Talent program," known as the Talenti project. This program, introduced at PoliTo, focuses on fostering excellence and inclusivity within the engineering domain [4].

The Talenti project emerged from the recognition that some students possess exceptional aptitude and potential in engineering. These students demonstrate a strong inclination towards STEM subjects and exhibit remarkable abilities compared to their peers, as determined by their TIL scores. The need to provide an enriched learning experience to these talented students arises from the desire to foster their intellectual growth, nurture their talents, and cultivate their potential as future leaders in the engineering field. This section will explore the rationale behind working with talented students and address the research questions that drive the Talenti project.

The Talenti project was initiated in the academic year 2014/2015 to provide specialized support and opportunities to top students pursuing engineering degrees. The project encompasses a comprehensive framework integrating talented students into the regular curriculum while offering additional resources and tailored activities to enhance their educational experience. The selection process is a critical component of the Talenti project, which aims to identify exceptionally bright students with the potential to excel in engineering. By analyzing the PoliTo incoming population over the past few years and their TIL scores, a distinct cohort of bright students has been observed, spanning all different engineering majors. Therefore, the project selection is based on the strong correlation between the students' scores in the TIL admission test and their subsequent academic performance, establishing the TIL score as a reliable selection instrument [3].

Approximately 7% of the students, out of the approximately 5,000 who undertake the TIL test every year, achieve a score equal to or greater than 70/100, making them the ideal candidates for the Talenti project. These students form a highly promising pool of participants. It is important to note that the initial experiment was conducted in Italian. However, the program is open to both national and international students, provided they have a basic knowledge of the Italian language. In any case, the standard courses within the program can be pursued in either Italian or English.

The number of participants in the Talenti project amounts to approximately 4% of the total student enrollment, corresponding to around 200 seats. To ensure continued participation in the program, students must meet specific checkpoints at the end of each academic year, including the end of the first semester, first year, second year, and third year. If students fail to fulfill the requirements at any of these checkpoints, they continue their studies through the regular academic path. Moreover, new students can be included in the program during the first two checkpoints while still adhering to the maximum enrollment capacity of 200 students.

The Talenti program offers talented students a structured and enriched curriculum, encompassing a range of activities and opportunities designed to stimulate their intellectual growth and foster their talents. The project implements a comprehensive and structured curriculum for the entire bachelor (three years), combining standard academic courses with dedicated activities. The program aims to provide a hybrid learning experience that extends beyond technical subjects, fostering students' development in nontechnical areas. The detailed structure of the program is depicted in Fig. 2.6.

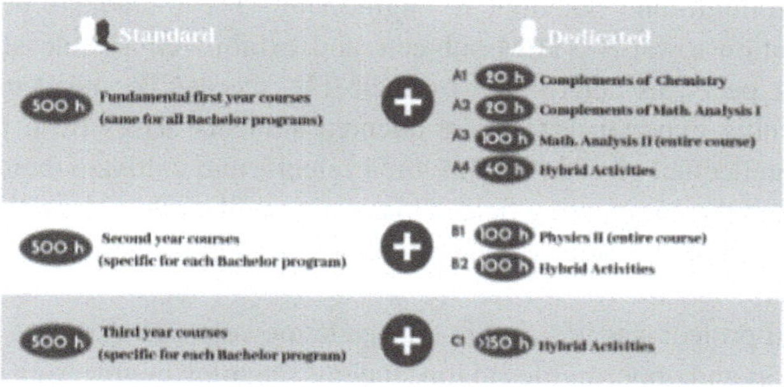

Fig. 2.6 The detailed structure of the program

During the first year, the selected students attend standard courses with their peers ("Anno comune"). These courses are conducted in parallel classes of approximately 250 students each. To enhance the student's learning experience, reinforcement activities are organized in the first semester for Chemistry and Mathematical Analysis I. These interventions promote practical application in laboratory sections, offer in-depth exploration of the covered topics, and encourage autonomous study. Interdisciplinary themes are also emphasized. In the second semester, ahead of the regular schedule, students participate in a reinforced course of Mathematical Analysis II, as well as Physics II in the third semester.

Hybrid activities form a core element of the Talenti project and are incorporated throughout the three-year curriculum. The program progressively increases the scope of these activities each year. Students attend a three-day full-immersion training weekend in the first year in November. This activity involves seminar classes to foster a creative problem-solving approach, team-based problem-solving competitions, and educational hiking in natural surroundings. This strategically positioned event allows students to familiarize themselves with one another and cultivate teamwork skills. During the second semester, a guided visit to CERN in Geneva is organized, providing students with an opportunity to meet scientists, explore the innovative research center's organization and management, and gain exposure to cutting-edge research.

In the second year, the Talenti project strengthened its ties with local industrial stakeholders through visits to renowned companies such as Maserati manufacturing plants. Additionally, laboratory sections are conducted within the FCA (FIAT Chrysler Automobiles) factory in Turin. These small-group laboratory activities enable students to reinforce their knowledge of labor processes. The experience comprises a two-hour plenary session focused on vehicle development, production processes, and automotive research objectives, followed by thematic sessions of 4–8 hours in CRF (Research Centre of FIAT) laboratories. Students can choose from various practical experiences, including crash finite element method analysis, computational fluid dynamics, electromagnetic compatibility, powertrain development, smart materials, and cooperative and wearable robotics. This hands-on engagement allows students to delve into specific phases of the research and development process, collaborating closely with FCA and CRF engineers and technicians.

In the third year, students are divided into small groups to pursue different activities. These activities include study abroad opportunities, encouraging students to spend a semester at partner universities within an extensive network of agreements, interdisciplinary projects focusing on improving the local environment in collaboration with public administrations, seasonal schools offering immersive learning experiences to strengthen specific knowledge, and internships providing work experience within innovative local factories. Furthermore, special events and conferences, such as the "Building up the future" lecture series, explore scientific, humanistic, and ethical aspects relevant to engineering.

Throughout the program, students benefit from the guidance of mentors who not only enhance their technical knowledge but also contribute to their cultural background, personal growth, and integration within the group. Mentors transfer technical expertise, professional experience, and labor market skills.

In addition, participating students receive various benefits, including educational incentives, economic advantages, and welfare support. To obtain the Talenti project diploma, students must fulfill all program requirements, complete all the required activities, and graduate within the specified timeframe (by December of the third year of the bachelor's program). Tuition fees are determined based on family income, with exemptions for students whose fees are less than or equal to 1,500 euros and a reduction of 1,500 euros for those with higher fees. Additionally, students receive a package that includes a local transportation subscription for bus and bike sharing and a card granting free access to numerous regional museums.

The Talenti project has yielded valuable findings and insights regarding the impact of working with talented students in EE. Analysis of the data collected from multiple cohorts reveals several noteworthy findings in the project's first four years.

Firstly, the gender distribution among the enrolled students was skewed, with 86% males and 14% females. Geographically, approximately 30% of the students came from Piedmont, the region where PoliTo is located. While 15% came from Apulia and 12% from Sicily, the remaining students represented other Italian regions and foreign countries. Evaluation of the students' TIL scores demonstrated a positive trend over time, indicating an improvement in the quality of enrolled students (Fig. 2.7). This trend is supported by an increasing number of first-year students achieving TIL scores higher than 70/100. Furthermore, the percentage of students graduating within three years or within 3.5 years has shown a consistent increment. A stable number of students in the program is observed across the first three cohorts, with the first cohort seeing 83% of Talenti students obtaining the "Talenti" diploma (Fig. 2.8). Regarding further studies, all students enrolled in the

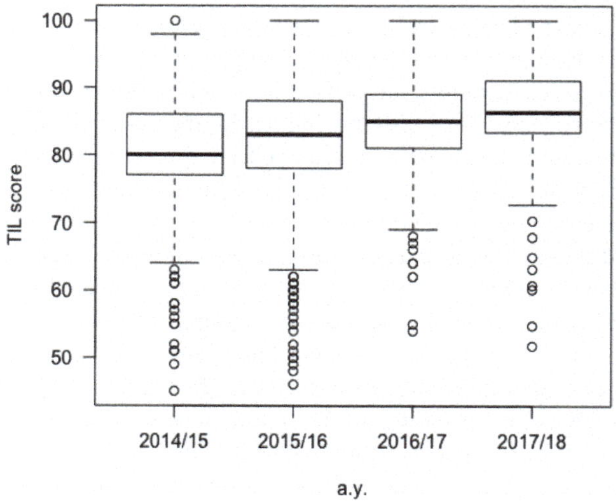

Fig. 2.7 Boxplot of the TIL score related to students of each cohort

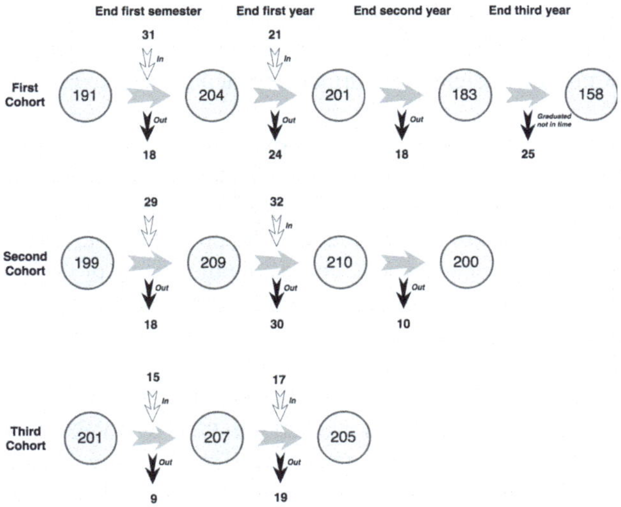

Fig. 2.8 Number of students involved step-by-step (in the circles), entering (white arrows) or leaving (black arrows) the program

project continued their education, with 124 of the 158 graduates applying to a master of PoliTo and 34 students pursuing their studies in highly ranked European universities. The qualitative analysis of students' feedback highlighted various aspects. For instance, mentorship support received low evaluation scores, leading to its deactivation and the introduction of a generic tutor. Additionally, students emphasized the importance of laboratory sections in Chemistry and suggested a better balance of workloads between semesters. As a result, adjustments were made, such as enriching Chemistry laboratories and moving the Computer Science course to the first semester. Trials and student feedback helped refine courses, leading to the introduction of a new Linear Algebra and Geometry course in the standard track (Fig. 2.9). The usage of benefits like the museum card, bike card, and public

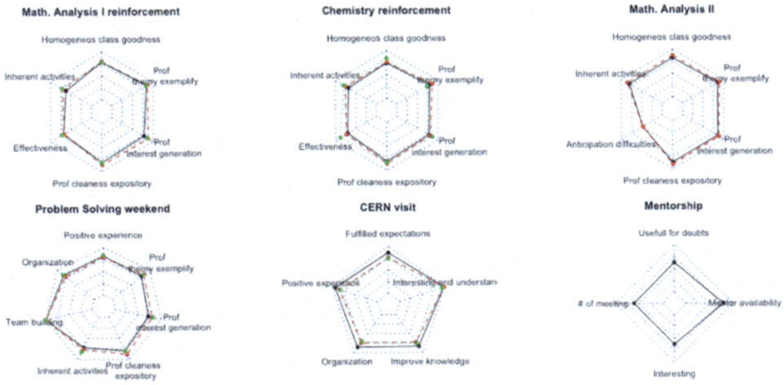

Fig. 2.9 Radar graph for each activity of the first year for the three cohorts

transportation card was found to be highly valuable for students. These findings contribute to understanding the Talenti project's effectiveness in improving student outcomes and provide insights for further enhancing the program's structure and offerings.

In conclusion, the Talenti project has proven to be a valuable initiative in EE, addressing the needs of exceptionally talented students and providing them with the resources and support necessary to reach their full potential. Through a rigorous selection process and a structured program, the project has fostered an inclusive and stimulating learning environment, cultivating the talents of the next generation of engineers. The findings of the Talenti project contribute to the broader field of EE, emphasizing the importance of catering to the needs of talented students and promoting collaborative learning for all students.

The "Percorso per i Giovani Talenti," in line with the objectives of the 2018–2024 strategic plan "PoliTo 4 Impact," has evolved into "INTRAPRENDENTI–Percorso per Ingegneria, Architettura, Pianificazione, Design Innovativo" starting from the academic year 2020/2021. This is a highly innovative three-year educational pathway parallel to the traditional course of study. It now includes not only 200 students from the engineering area but also extends to 40 students from the architecture area (including architecture, urban and regional planning, and design and communication), selected based on their TIL scores.

2.5 Reverse Inclusion: A New Perspective

One of the key strengths of the Talenti project lies in its capacity to stimulate and motivate not only the talented students themselves but also the broader student community [6]. The project establishes an environment that fosters peer learning and collaboration by incorporating talented students into regular classes. This approach yields positive outcomes for students overall, including reduced dropout rates, improved academic performance, and increased graduation rates.

Reverse inclusion, a concept closely associated with the Talenti project focuses on the indirect inclusion and support of protégé students through the participation of talented students in regular classes. Speaking about "reverse inclusion" can seem a paradox. However, the empirical definition can be stated as the process that, by stimulating the talented students, indirectly includes the protégé. This idea came from the observation of the impact that the presence of bright students attending the Talenti project can have a beneficial impact on the learning environment for all students.

By integrating talented students in standard class settings, reverse inclusion promotes peer interaction, knowledge sharing, and exchange of ideas. In these mixed-class settings, talented students serve as role models and mentors, guiding and supporting their peers. This interactive environment allows students with varying ability levels to engage with challenging subject matter and cultivate a deeper comprehension of the material. By working collaboratively on projects

Table 2.5 The number of students who graduated in 3 years and within 3.5 years with the related average degree

a.y.	Matriculated	Graduated in 3 years	Average degree 3 years	Graduated within 3.5 years	Average degree 3.5 years
2011/2012	3989	733	100,4/110	1078	94,2/110
2012/2013	4007	872	100,0/110	1152	96,8/110
2013/2014	4168	893	100,9/110	1261	97,1/110
2014/2015	4432	980	101,2/110	1390	100,1/110

and assignments, students can benefit from the diverse perspectives and expertise brought by the talented students. Furthermore, the participation of talented students in regular classes provides an opportunity for protégé students to observe and learn from the accomplishments of their peers, thereby fostering motivation and raising the overall quality of education. This has been registered with a reduction in the graduation time with a higher score degree average, despite an increase of first-year students (Table 2.5).

The reverse inclusion approach also facilitates curriculum improvements. As talented students represent a standard class size, they can offer valuable feedback on course content, organization, and workload distribution. This feedback loop enables lecturers to optimize course design, identify areas for improvement, and tailor the curriculum to meet the needs and capacities of all students. The insights gained from talented students' experiences can inform pedagogical strategies, assessment methods, and the integration of real-world applications into the curriculum.

For example, as described in Sect. 2.4, based on the annual survey findings, talented students expressed concerns regarding the workload imbalance between the first year's two semesters. To address this issue, the Computer Science course was identified as the most suitable candidate for being moved from the second to the first semester, considering the students' background requirements. Consequently, starting from the academic year 2015/2016, this course adjustment was implemented for all students, ensuring a more balanced curriculum.

Similarly, the standard Geometry class required a content review focusing on reinforcing the Linear Algebra component and introducing the use of Matlab software. A new course combining Linear Algebra and Geometry was meticulously planned. Leveraging the talented students' ability to simulate a standard class environment and provide valuable input, the revised course was proposed within the Talenti program. As their feedback indicates, the students responded positively to this new course (Fig. 2.10). This trial not only facilitated the coordination between different courses but also led to the decision to introduce the new "Linear Algebra and Geometry" course to all students following the traditional path, starting from the academic year 2016/2017, thereby replacing the previous Geometry course.

The concept of reverse inclusion carries significant implications for EE, as evidenced by the "Inclusion: a new reverse perspective" [6] paper's recognition with

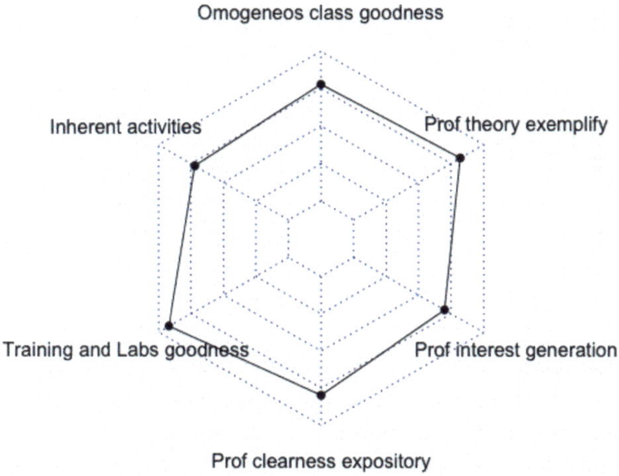

Fig. 2.10 Radar plot with the results of the Linear Algebra and Geometry course's survey

the "Most Original Contribution" award at the UK and IE EERN symposium on November 2nd, 2018, in Portsmouth.

By integrating the Talenti project and reverse inclusion strategies into EE, institutions can establish inclusive learning environments catering to student's diverse needs and abilities. This integration necessitates designing inclusive programs that address the challenges and opportunities of nurturing talented students in regular classroom settings. Additionally, providing appropriate resources and support for talented students, such as advanced learning materials, specialized workshops, and mentorship programs, further enhances their educational experience. Furthermore, fostering meaningful interactions among students with varying abilities through group projects, peer mentoring, and collaborative problem-solving promotes a culture of inclusivity and collective learning.

In summary, the Talenti project and the concept of reverse inclusion challenge the traditional teacher-centered model and promote student-centered learning environments that foster collaboration, critical thinking, and the development of higher-order skills. By embracing these approaches, PoliTo has created an educational landscape where the potential of all students, including talented individuals, is recognized, nurtured, and harnessed to benefit the entire student body.

2.6 Methodological Support for Learning How to Study: DayByDay

The DayByDay project was developed based on a meticulous analysis of the *Analisi Matematica 1* (Mathematical Analysis 1) course, employing the principles of the ADDIE (analysis, design, development, implementation, evaluation) instructional design methodology (for further details see Sect. 3.2.1). The role of introductory engineering courses, such as Mathematical Analysis 1, is not limited to pure knowledge transfer but represents a preliminary approach to engineering. As a secondary goal, they also reinforce many soft skills that engineering studies require.

Incorporating the ADDIE methodology ensured a systematic approach to developing and implementing the DayByDay project. In the analysis phase, the lecturers, guided by the EER, conducted a comprehensive needs assessment, examining the students' performance in the TIL assessments and analyzing the data from their academic careers. The TIL results indicated that a considerable number of students struggled with specific mathematical concepts and needed help applying them in problem-solving scenarios. The analysis of the students' careers further revealed that they faced methodological challenges related to study habits, time management, and self-assessment, which impacted their graduation time and the number of exams passed in each session.

In the academic year 2019/2020, a pilot qualitative study was conducted to support the project's development. This study involved only one of the twenty parallel courses, with approximately 60 out of 250 students in the course participating voluntarily. These students did not receive additional points for the final exam, but they received a detailed study program and the opportunity to take an online weekly quiz. The students received email reminders if the quiz was not completed within a week. At the end of the semester, the data relating to the weekly quizzes and the exam outcomes were analyzed. The majority of voluntary students who participated in the pilot study successfully passed the exam in the winter session. This positive outcome provided valuable insights and served as the basis for the structured DayByDay project that was designed and implemented in the academic year 2020/2021.

During the academic year 2019/2020, a group of dedicated lecturers regularly came together to follow the ADDIE methodology. It thoroughly examined the results from the TIL assessments, the reverse inclusion study, and the pilot study. This rigorous analysis not only revealed content-related gaps but also shed light on the methodological challenges faced by the students.

Using these valuable insights, the lecturers proceeded to the design phase of the DayByDay project. They devised strategies to address the identified challenges by leveraging appropriate rewards and instructional resources to support both the content and methodology. They decided to introduce ongoing activities to support the continuous engagement of students, and these activities were assessed and counted towards the final exam score, with a maximum weight of 3 points. This change required revising the entire exam structure, as summarized in Table 2.6.

Through a careful selection process, the lecturers created a set of activities and materials that aimed to scaffold the student's learning process effectively. They also took into account the unique circumstances imposed by the COVID-19 pandemic, incorporating measures to support distance learning and foster student engagement.

To study the impact of this new approach, the DayByDay project was made available to 7 out of the 20 parallel mathematical courses at PoliTo. Each course consisted of 60 hours of lectures and 40 hours of exercise classes. The theoretical lessons

Table 2.6 Structure of the Mathematical Analysis 1 before and after the introduction of the ongoing activities in the academic year 2020/2021

	a.a. 2019/2020	a.a. 2020/2021	
Multiple-choice test (T)	Min 12/20	Min 8/15	Same for all courses
Written exam (W)	Min 5/13	Min 8/15	Same for all courses
On-going (Og)	–	Max 3	Each course can organize its own
Oral exam (O)	Not mandatory	Not mandatory	
Final grade	T+W+O	T+W+Og+O	

Table 2.7 Evaluation criteria in the DayByDay project

Evaluation criteria	Description	Mode	Max activity points for each
Individual activities [T]	4 sets of 15 multiple-choice tests [T1, T2, T3, T4].	1 attempt only, 1 hour, open a week	T: 15
Group activities [E and EC]	vf 3 sets of 12 exercises [E1, E2, E3] to be performed with teammates. Then, each team is asked to evaluate the work of another team [EC1, EC2, EC3].	1 week to submit the set of exercises, 1 week to evaluate the work of another team	E+EC: 36 (24+12)

were devoted to presenting the topics, including definitions, theorems, examples, properties, and proofs, which were believed to facilitate the learning process and students' metacognition. Every theoretical aspect was associated with introductory examples. The exercise hours aimed to develop proficiency in computation.

As the project entered the development phase, the lecturers refined its structure to ensure clarity in communication and prompt student support. They established well-defined evaluation criteria, allowing students to assess their own progress and receive constructive feedback. An overview of the evaluation criteria for the main activities proposed by the DayByDay project is presented in Table 2.7.

An activity was proposed weekly, following the pattern T, E, EC (as shown in Fig. 2.11, the activity calendar for the academic year 2020/2021). Students could choose when to fulfill these activities within the week, starting from the fourth week. All the activities replicated the exam environment, with the tests using the same structure and platform and the exercises structured similarly to the written part.

The individual activities consisted of four sets of 15 multiple-choice tests performed via the Moodle platform. Each test allowed only one attempt and lasted one hour. The group activities involved three sets of 12 structured exercises to be solved in collaboration with teammates. The solutions were uploaded to Moodle. The following week, each team received another group's solution and was asked to evaluate it, correct any mistakes, and upload the revised document on Moodle. Students received a structured layout where they needed to fill in the list of participants, the exercise solution, the peer-to-peer score, and the corrections made by the lecturer. The lecturer could confirm or modify the peer-to-peer score associated with the exercise solution and award one point for an acceptable

data	Inizio attività **ore 16:00**	Consegna attività **entro ore 12:00**
30 ottobre 2020	**T1**	
6 novembre 2020	**E1**	**T1**
13 novembre 2020	**EC1**	**E1**
20 novembre 2020	**T2**	**EC1**
27 novembre 2020		**T2**
4 dicembre 2020	**E2**	
11 dicembre 2020	**EC2+T3**	**E2**
22 dicembre 2020	**E3**	**EC2+T3**
4 gennaio 2020	**EC3**	**E3**
8 gennaio 2021	**T4**	
11 gennaio 2021		**EC3**
15 gennaio 2021		**T4**

Fig. 2.11 Activities calendar for academic year 2020/21

correction or zero if something were missing or incorrect. Each group activity could receive a score of up to 36 (24 for the solution and 12 for the correction).

The evaluation criteria for the activities were as follows:

- Individual activities:
  - 1 point: Four tests performed with a score of 8/15 or higher, with at least two tests scoring 12/15 or higher.
  - 0.5 points: At least three tests were performed with a score of 8/15 or higher.
- Group activities:
  - 2 points: Three sets of exercises delivered with a score of 20/24 or higher, with at least two sets scoring an overall score of 32/36 or higher after the correction phase.
  - 1.5 points: Three sets of exercises delivered with a score of 20/24 or higher, with at least one set scoring an overall score of 32/36 or higher after the correction phase.
  - 1 point: At least two sets of exercises delivered with a score of 20/24 or higher.
  - 0.5 points: At least one set of exercises delivered with a score of 20/24 or higher.

During the implementation phase, the DayByDay project was successfully executed. The lecturers ensured clear and timely communication with the students, providing them with a structured learning environment. They offered well-defined evaluation criteria, enabling students to monitor their progress effectively. Moreover, the lecturers remained attentive to the potential stressors and difficulties faced by students, offering necessary support with the help of three senior tutors.

At the beginning of the course, the lecturers organized homogeneous groups of about ten people based on the admission test results. This allowed for maintaining similar average scores and variance within each group. Each group autonomously

Table 2.8 Number of students in each course within DayByDay with the relative participation in the activities

			T		E&EC	
# Course	Students	New enrolled students	Min	Max	Min	Max
1	268	213	182	217	213	217
2	402	329	285	315	315	344
3	250	205	172	199	193	206
4	261	224	192	217	208	216
5	334	211	204	234	228	236
6	252	208	175	212	204	218
7	59	56	49	55	55	55
Total	1826	1446	1259	1449	1416	1492

chose a spokesperson responsible for file uploads and coordinating meetings for the group activities. Students could decide where to meet, but all meetings were held online due to the pandemic conditions. The lecturer had the option to join these meetings to check participation and the distribution of the workload within the team. The lecturer played an auditing role and did not intervene in the discussions.

The number of students participating in the DayByDay project and their engagement levels are presented in Table 2.8.

An evaluation of the project's effectiveness was conducted to assess its impact. A comparative analysis between the intervention group (the seven parallel courses that adopted DayByDay) and the control group revealed significant differences in student outcomes. One crucial element for evaluating the impact of the DayByDay project is the number of students who decide to take the exam in January. Applying to the first available call at the end of the course implies that students have followed the course and remained on track with their studies. In the academic year 2020/2021, the experimental group had a 2.34% higher number of students who sat for the test than the control group (Table 2.9).

However, it should be noted that for both groups, the overall number of examinees decreased, which can be attributed to the pandemic conditions and the shift to online courses. The two groups were comparable in the number of students who passed the multiple-choice test in the academic year 2019/2020. The test had random questions, but each set was comparable in difficulty and the variety of topics addressed. Therefore, the test provided an objective measurement. In the

Table 2.9 Test results related to January call in the academic year 2020/2021

	January 2020		January 2021	
	Experimental group	Control group	Experimental group	Control group
Test passed	730 (47,04%)	1560 (48,84%)	839 (56,38%)	1587 (50,64%)
Test not passed	548 (35,31%)	1121 (35,10%)	313 (21,04%)	766 (24,44%)
Did not show up	274 (17,65%)	513 (16,06%)	336 (22,58%)	781 (24,92%)
Total	1552	3194	1488	3134

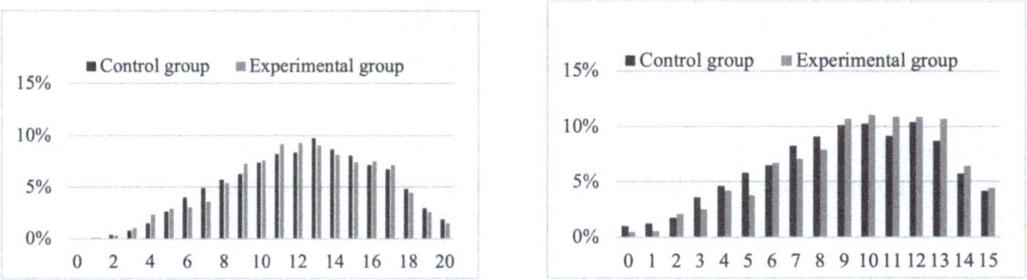

(a) (b)

Fig. 2.12 Test score distribution in (**a**) January call 2020 and (**b**) January call 2021

academic year 2020/2021, the experimental group had a 5.74% higher number of students who passed the test compared to the control group. Analyzing the score distribution (Fig. 2.12b), the experimental group had a higher average (9.51/15 points vs. 9.05/15 points) with a lower standard deviation (3.37 vs. 3.53). Similarly, in the academic year 2019/2020, there was no significant difference in the scoring average (12.12/20 vs. 12.22/20) and standard deviation (4.02 vs. 4.03) between the two groups (Fig. 2.12a).

In the end, considering the overall exam, which includes the written test that is the same for all courses, the assessment becomes less objective compared to the test. In a tiny proportion of cases, the lecturers or the students required an oral exam. Table 2.10 shows the results of Analysis Mathematical 1 at the end of the first session (including both the January and February calls).

The results indicate the positive impact of the DayByDay project. The intervention group exhibited higher engagement and success rates than the control group. Furthermore, the intervention group achieved significantly higher average exam scores, demonstrating a deeper understanding of the course material. These results have been presented as a poster at the SEFI conference [1]

In conclusion, the DayByDay project exemplifies the value of methodological support in EE. By applying the principles of the ADDIE instructional design methodology and leveraging insights from the TIL assessments and the reverse inclusion study, the lecturers developed a comprehensive framework that effectively addressed content-related gaps and methodological challenges. The project's outcomes, as evidenced by the evaluation results, underscore its effectiveness in enhancing student learning experiences and improving academic performance.

Table 2.10 Exam results related to the first session

	First session 2020		First session 2021	
	Experimental group	Control group	Experimental group	Control group
Exam passed	851 (54,84%)	1662 (52,03%)	823 (55,31%)	1441 (45,98%)
Exam not passed	486 (31,31%)	1135 (35,54%)	431 (28,96%)	1168 (37,27%)
Did not show up	215 (13,85%)	397 (12,43%)	234 (15,73%)	525 (16,75%)
Total	1552	3194	1488	3134

2.7 Math Game Without Frontiers: Reinforcing Mathematics Through Playful Learning

In addition to the DayByDay project, another innovative initiative called the Math Game Without Frontiers [12] was introduced to enhance further the learning experience for first-year students in the Mathematical Analysis 1 course. This initiative aimed to reinforce the course contents through engaging and interactive games, incorporating playful learning elements to provide a unique and enjoyable approach to mastering mathematical concepts.

The Math Game Without Frontiers served as a complementary tool to traditional classroom instruction and study materials, creating a dynamic learning environment that fostered active participation and critical thinking among students. Through various game-based activities, students were able to apply their mathematical knowledge in practical and stimulating contexts, enhancing their understanding and retention of the course material.

This section delves into the details of the Math Game Without Frontiers competition, exploring its objectives, methodology, and the positive outcomes it yielded. By incorporating playful experiences into the learning process, this initiative aimed to make studying the first Mathematical course at the university more engaging and enjoyable for first-year students.

The Math Game Without Frontiers competition took place at the College of Merit CAMPLUS, located in different cities in Italy from south to north, during the autumn of 2020 and 2021, involving first-year STEM students. Initially held in person, the competition transitioned to an online format due to the COVID-19 pandemic, enabling participation from colleges all over Italy. The competition consisted of several rounds, with teams of students competing against each other to earn points based on their performance in the mini-games.

The challenge was open to all first-year students living in one of the Italian Camplus and with at least one mathematical exam in the university's first year. The first edition involved 52 participants (15 females and 37 males) from 9 colleges. In the second edition, there were 30 participants (14 females and 16 males) from 7 colleges. Students autonomously formed couples within the same Camplus, with some exceptions in the first edition when two mixed pairs came from different Camplus. Teams were free to choose whether to play physically or remotely. A private instantaneous message channel between lecturers and each couple was established to collect the game's solutions.

Applying the game-based learning methodology, the games incorporated well-known puzzles to familiarize students with Mathematical basic content, partially inspired by the work of Morando [40] and further developed for remote play. Math Games Without Frontier consisted of five one-hour-long sessions of playful scored activities. The events took place one week apart and required mathematical knowledge, teamwork, and strategy. In the 2020 edition, the first three evenings focused on essential knowledge (elementary functions such as trigonometric, exponential, and logarithmic functions; equations and inequalities), while the last two reinforced the concepts of limits and derivatives. In the 2021 edition, the first two evenings work on essential knowledge, while the remaining three reinforce the concepts of limits, derivatives, and integrals. The inclusion of questions about integration in the second edition was based on feedback from the final survey of the competition's first edition. The rules and any printed materials for each challenge were made available to students the day before each session. Each evening consisted of three different games played in various rounds. To maintain students' concentration throughout the game session, the lecturers showed and discussed all the solutions at the end of the meeting. Teams received their scores during the week, and their ranking positions were provided at the beginning of each session. At the end of the experience, a winning team and a winning Camplus were selected, each receiving a prize.

The "Math Game Without Frontiers" competition employed several mini-games to challenge and assess students' mathematical skills. These games included Labyrinth, Ordering, Math Twins, Colouring Puzzle, Treasure Hunt, Guess What?, Target, Sudoku, and Connect the Dots. Each game focused on different mathematical concepts, requiring students to apply their knowledge and problem-solving abilities within a limited timeframe. Each game's rules and scoring criteria were carefully designed to promote engagement and competition among the participants. Here, we provide a summary of the basic rules used during the competition for each game:

Labyrinth

The Labyrinth game presented True and False exercises in which the tutor read ten statements twice. Starting from the center, the answers formed a path to exit the labyrinth, with true answers matching the green star and false answers matching the red star (see Fig. 2.13a). After the reading ended and 30 seconds elapsed, all the statements appeared on the screen for 1 minute, with the game concluding 30 seconds later (a total of 120 seconds from the end of the reading). Teams earned 1 point for each correct answer, two bonus points if they submitted within 30 seconds of the reading's end, and one bonus point if they submitted within 90 seconds. The first team to submit gained an additional two bonus points. This exercise not only assessed mathematical skills but also trained listening abilities that are not typically associated with Mathematics.

Ordering

The game is divided into two rounds: in the first one, teams have to order eight values of trigonometric functions in ascending order, while in the second, eight values must be sorted in a decreasing manner. Each heat has a duration of 4 minutes. Each team has a "STOP" to indicate the point to end the sequence's validation.

(a) (b) (c)

Fig. 2.13 Solutions of different games. (**a**) Labyrinth. (**b**) Coloring. (**c**) Treasure hunt with the target) [image: freepik.com]

Teams have one extra minute for the sequence's delivery. Pairs gain 1 point for each correctly ordered sequence but get 0 points if the series order is wrong, even for a single value. Teams get two bonus points if the complete sequence is correct and delivered before time runs out.

Math Twins

A table of 12 + 12 elements to be paired is given to each team. In 5 minutes, the team has to form as many pairs as possible, and within 1 minute, they must send the complete list of pairs. Each team gains 1 point for each correct pair but loses 1 point for each wrong couple. If all pairs are matched correctly, two bonus points are awarded. If the total score is negative, the team score equals zero.

Colouring Puzzle

Another example is the Colouring Puzzle, where each team had to blacken spaces in a figure containing the solutions of 15 mathematical expressions (such as inequalities, equations, or limits; see Fig. 2.13b). The available time for this activity was 12 minutes, and once completed, the team had to submit a photo of the obtained image to the tutors. Suppose the figure is colored correctly; the couple scores ten points. The team loses 1 point for each incorrect or missing box. The team gains two bonus points if the correct figure is delivered before time runs.

Treasure Hunt

Each team has to solve a system of inequalities graphically, with the goal of identifying the object constrained in the solution plane region of the system. Axes origin is assumed in the lower-left corner (see Fig. 2.13c). The available time to complete the game is 5 minutes. Teams that deliver the correct object within 3 minutes gain ten points; couples that provide the proper object after 3 minutes earn five points. If the object is wrong, the team gets no points. Two bonus points are assigned to the first team that identifies the correct object.

Guess What?

Thirty plots of functions are shown to teams. The tutor secretly chooses one between all the figures. Couples can ask questions about a specific feature defined in advance, with yes or no answers, to guess the tutor's choice. Teams that think they have identified the correct function send a message to the tutors, and the game ends as soon as a pair correctly guesses. Each team has at most three attempts, playing the game in multiple rounds. The team that identifies the correct function gains three points; the same team gains one bonus point if it identifies an admissible analytic expression within 30 seconds and two bonus points if it identifies two different expressions in 30 + 30 seconds.

Target

A target made up of boxes is shown to teams. Each box contains one function. The game's goal is to reach the center of the target by passing over each box at most once, starting from the pointed box. One can move from one box to another if the two boxes are adjacent and the functions follow a prescribed property (for example, Fig. 2.14a shows the case in which the rule is: the intervals of increasing monotony of two functions have a non-empty intersection). The available time is 7 minutes plus one extra minute for delivery. Teams gain 1 point for each correct passage between boxes. They earn one bonus point if they deliver within 7 minutes and gain another two bonus points if, for each box, they indicate the specific property (for example, the interval in which the function is increasing).

Sudoku

Each team has to solve a Sudoku with some empty colored dots (see Fig. 2.14b). Each color is associated with a definite integral. To solve them, they must correctly place the numbers and complete the Sudoku. The time to complete the game is 20 minutes, plus one minute for delivery. Teams are asked to deliver the starting scheme with the integrals computed in order to obtain points: one point for each correct result within 10 minutes and 0.5 points for each proper answer delivered

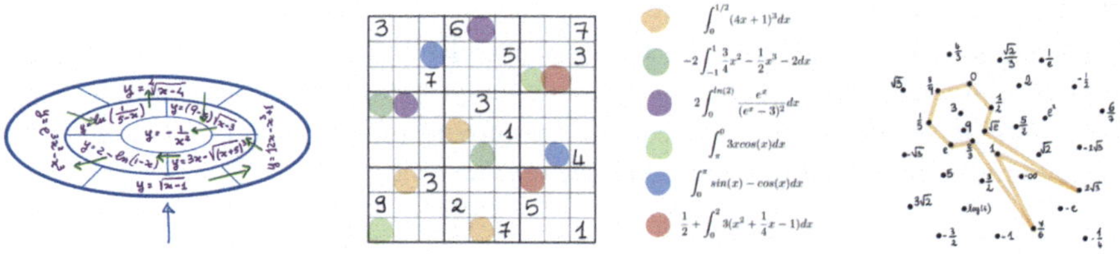

(a) (b) (c)

Fig. 2.14 Examples of different games. (**a**) Target. (**b**) Sudoku. (**c**) Connect the dots

before the end. Six bonus points are gained by the teams that complete the Sudoku within 15 minutes, five points within 20 minutes, and 4 points if the solved scheme is delivered at the end of the time.

Connect the Dots

Each team must solve ten limits and, following the given order, use the solutions to connect the related dots (see Fig. 2.14c). The aim is to draw the secret image within 10 minutes. Each number correctly linked gives one point. The team receives two bonus points if they deliver the correct picture within 10 minutes.

The "Math Game Without Frontiers" competition has positively impacted the participants' mathematical abilities. Students improved their logical reasoning, calculation skills, and graphical understanding of mathematical concepts through engaging in the mini-games. The competition also provided teamwork, strategy development, and healthy competition opportunities. Feedback from the participants indicates a high level of satisfaction with the experience, with many students expressing a desire for future competition editions. Moreover, the games were found helpful and educational in aiding students' understanding of mathematical concepts. The limited time constraints fostered healthy competition, making Mathematics more challenging.

Furthermore, analyzing each game's results provided insights into Mathematics' critical aspects for first-year students. For instance, the Labyrinth game was played on three different evenings, and initial concentration was identified as the main difficulty when answering True or False questions based on spoken sentences. As students are primarily accustomed to solving mathematical problems by reading the written text, in both editions, only a few teams delivered their solutions on the first evening before instructions appeared on the screen. However, on the last evening, the number of couples who immediately returned after the end of the reading increased considerably, indicating an improvement in their skills to formalize problems spoken aloud.

The Ordering game was played only on the first evening but also highlighted a critical fact discovered in other games. Students need to familiarize themselves with trigonometric expressions and find it difficult to quantify them. In 2020, 75% of teams scored zero despite having the "stop" available. Similar results came out in the 2021 edition, where 50% of couples were unable to solve at least one of the two "ordering" games proposed. This suggests that many students still feel confident in the results, although they still need to gain a complete mastery of trigonometric functions. Therefore, this could explain why few have used the stop.

The Math Twins game confirmed students' difficulty associating the algebraic expression of a function with its graphs. The game was played on three different evenings, and each evening, there was a different type of twins: trigonometric expressions and numerical values, functions and derivatives, functions and graphs. Teams reached higher scores in the game when the associations were not graphical. Conversely, scores are lower when students associate the graph with the analytic expression; notice that almost 50% of teams scored zero on the first evening of both editions. Scores of the other games suggest the same conclusions. In particular, the Treasure Hunt game highlighted again that most students need to become more familiar with the graphical exercise's solution.

Furthermore, a common factor in all games is the short time to complete challenges: Quickly solving exercises requires a specific mastery of Mathematics. Since the goals of this challenge are to develop the ability to intuitively visualize the graph of an analytic function and support preparation in differential calculus, these games suggest that more emphasis has to be devoted to the graphical part.

In the second edition, we introduced two new games: Sudoku and Connect the Dots. Regarding the Sudoku, only three teams were able to solve the integrals and complete the scheme. It seems that students made mistakes in calculating elementary integrals quickly. Then, when they got stuck on the Sudoku, they did not realize that the error was on the computation of the integrals, not the scheme solution. Similar considerations can be drawn for the second added game. As before, the time limit seems to be a significant obstacle for students.

The Math Game Without Frontiers competition has not only provided an enjoyable and interactive learning experience for the students. Still, it has also effectively complemented the teaching of the first university course in Mathematics, enhancing students' mathematical skills and preparing them for further studies in the field.

2.8 STEMM Games: Developing Soft Skills While Playing

The environment of merit colleges, as exemplified by the "Math Games Without Frontiers" competition, provides a fertile ground for exploring cutting-edge multidisciplinary courses, thanks to the diverse composition of its students. At PoliTo, students have the opportunity to access two merit colleges in the region: Collegio CAMPLUS and Collegio R. Einaudi (present in multiple locations).

In this subsection, we will explore the concept of STEMM games and their role in developing soft skills among students. The experience has been presented at the IEEE EDUCON 2024 conference [2]. We will discuss a specific edition that took place at the Collegio R. Einaudi. Building upon the richness of multidisciplinarity present in these colleges, the project aimed to leverage the multidisciplinary nature of the academic community to design a course that follows the principles of playful learning and game-based learning, ultimately strengthening the students' soft skills.

STEMM games refer to educational games that revolve around science, technology, engineering, Mathematics, and medicine subjects. These games are designed to provide an interactive and engaging learning experience, combining gameplay elements with educational objectives. Unlike gamification, which involves integrating gamelike features into nongame environments, game-based learning employs actual games, often on digital platforms, for educational purposes with defined learning outcomes [44]. Games' immersive nature and social perspective make them an ideal medium for effective learning in higher education, as they increase enjoyment and reduce anxiety while enhancing deep learning and higher-order thinking [13, 21].

STEMM games offer a unique opportunity to foster soft skills development among students. Soft skills, also known as transferable or nontechnical skills, are essential attributes that complement technical knowledge and expertise. They include problem-solving, teamwork, time management, creative thinking, and critical thinking, among others. These skills are highly valued in the workplace and are crucial to personal and professional success.

The project involved a series of games conducted over five sessions, each focusing on a specific theme related to one of the five areas: Science, Technology, Engineering, Mathematics, and Medicine. The challenge duration was 11 hours, divided among the sessions, with 27 students participating, forming five teams. The project took place between October and mid-December 2021, and each group gathered physically in a separate space while being connected online with other groups and tutors.

The STEMM games kicked off with an engaging introduction night, setting the stage for the exciting journey that awaited the participating students. During this inaugural event, each student was invited to complete an online self-assessment questionnaire on their soft skills, utilizing the SoftSkillOrienta platform (https://softskillorienta.ricercaconlascuola.it/SkillMapping/Enter)[33]. The questionnaire generated a short report with a spider diagram Fig. 2.15, offering insights into the students' proficiencies across different areas. This initial evaluation not only provided a baseline understanding of their strengths but also served as a starting point for personal and collective growth throughout the project. The introduction night fostered a sense of anticipation and curiosity among the students, sparking their enthusiasm for the upcoming STEMM games and the opportunity to develop their soft skills in a dynamic and interactive learning environment.

The sessions were organized as follows:

Fig. 2.15 Example of the spider diagram inside an individual report

2.8.1 *Game 1: Engineering*

The first game focused on Engineering. It was organized in a structured manner to ensure an engaging and competitive experience for the participating teams. The challenge involved constructing towers with the goal of achieving maximum height while being capable of supporting the greatest weight. Each team was provided with materials such as A4 papers, large adhesive tapes, canned goods weighing 400 grams each, and a measuring tape to facilitate the competition.

The organization of the game consisted of three distinct phases. Firstly, there was a 15-minute strategy session where team members could discuss their approach with the camera turned on but microphones muted. This allowed teams to plan their tower construction strategies collaboratively. Next, there was a 30-minute construction phase, where team members actively worked on building their towers while keeping their cameras and microphones on for real-time communication and coordination. This phase provided an opportunity for teams to apply their problem-solving skills, teamwork, and leadership abilities.

Finally, there was a 20-minute construction phase where the camera was on, but the microphones were muted, allowing teams to focus on finalizing their towers without distractions. At the end of the designated construction time, the towers were evaluated based on two criteria: height and strength Fig. 2.16. Points were awarded based on the tower's height, with the tallest tower receiving 50 points. Additionally, the towers' resistance to holding canned goods for specific durations resulted in additional points awarded based on the number of cans supported and the duration of support.

This organization of the first game provided a structured framework for teams to strategize, collaborate, and put their engineering skills to the test. The planning, construction, and evaluation phases allowed participants to showcase their problem-solving abilities, teamwork, and creativity in designing and building sturdy towers.

Fig. 2.16 Picture of the towers before measuring the height

2.8.2 *Game 2: Medicine*

In collaboration with Dr. Chiara Vignati, the second game centered around Medicine. The challenge was meticulously organized to facilitate engaging debates and foster critical thinking among the participating teams. "Defend Your Position" followed a structured format that allowed teams to present their arguments, counter-arguments, and concluding remarks within specified timeframes.

Each team defended either a favorable or opposing position, and different team members served as speakers for each argument. The thesis presented in the medicine game challenged participants to explore the topic of inclusivity in athletics. Specifically, the question posed was whether Paralympic athletes who meet the participation criteria should be allowed to compete alongside nondisabled athletes in track and field events. This thought-provoking thesis encouraged students to delve into the ethical, logistical, and practical considerations surrounding inclusivity in sports.

The organization of the game involved several phases. Firstly, there was a 15-minute planning session where teams could strategize their arguments with their cameras turned on but microphones muted. This initial phase allowed teams to coordinate their approach, assign roles, and develop their key points.

Next, there were multiple rounds of argumentation, each with specific time allocations. In the first round, each team had three minutes for their speaker to present their arguments. Following this, there was a 10-minute planning phase with cameras on and microphones muted, allowing teams to prepare counter-arguments. In the second round, each team had two minutes for their second speaker to present their counter-arguments.

Subsequently, teams had another 10-minute planning phase to prepare their concluding remarks, once again with cameras on and microphones muted. In this last round, their position was forced to change: the team favoring the thesis would argue an opposing position and vice versa. Each team had one minute for their third speaker.

Throughout the game, points were awarded based on the clarity, persuasiveness, and organization of the arguments presented. The judges evaluated the quality of the arguments and the overall effectiveness of each team's communication skills. This well-structured organization of the second game provided a platform for teams to engage in thoughtful and well-articulated debates on various medical topics.

The timed rounds and planning phases allowed for a balanced exchange of ideas, while the evaluation criteria emphasized the importance of critical thinking, persuasive communication, and empathy. By examining the implications from various perspectives, participants had the opportunity to critically analyze and construct compelling arguments for or against the integration of Paralympic athletes into able-bodied competitions, developing and showcasing their skills in argumentation, critical analysis, and effective communication.

2.8.3 Game 3: Science

The third game of the STEMM project focused on the field of science and was thoughtfully organized to enhance scientific communication and foster collaboration among the participating teams. The game revolved around the challenge of becoming effective science communicators tasked with explaining a given scientific topic in the most straightforward possible manner while covering all key concepts. The game followed a well-defined structure to encourage teamwork, resilience, and trust.

The organization of the game involved several phases. Firstly, teams were given three different scientific topics and had 15 minutes for organization. During this planning phase, the teams could discuss their strategies and allocate roles for each member. Cameras were turned on, but microphones remained muted, allowing teams to collaborate visually while formulating their approach.

Next, there was a two-minute session where teams had available the keywords of two of the three topics assigned to them. This phase provided a glimpse into the chosen topics and allowed other teams to guess and engage in interactive learning. Following this, teams underwent a series of three rounds, each lasting five minutes.

In each round, a team member was randomly chosen as the presenter, who had one minute to explain the assigned topic most straightforwardly, focusing on key concepts. The team's performance was evaluated based on the clarity and simplicity of their explanations, as well as the number of correctly guessed keywords by other teams. This format fostered collaboration, resilience, and trust among team members as they collaborated to distill complex scientific concepts into accessible explanations.

The structured organization of the third game provided a platform for teams to enhance their collaboration, develop resilience in scientific communication, and cultivate skills in simplifying complex concepts. By engaging in these activities, participants not only expanded their scientific knowledge but also refined their ability to communicate effectively and engage with diverse audiences.

2.8.4 Game 4: Mathematics

The fourth game of the STEMM project, centered around Mathematics, was thoughtfully organized to foster problem-solving skills, autonomy, and creativity among the participating teams. The game encompassed two phases: problem-solving and the invention of a new game based on an assigned topic.

The organization of the game involved several steps. Firstly, teams were given 20 minutes to problem-solve with their cameras on but muted microphones. During this phase, teams collaborated to solve mathematical problems, leveraging their problem-solving skills, critical thinking abilities, and mathematical knowledge. They worked together to devise solutions, discuss strategies, and analyze the mathematical challenges. Like the one used in the "Math game without frontiers" (see Sect. 2.7).

Next, there was a 30-minute phase dedicated to inventing a new game. Teams brainstormed and created a unique game concept based on the assigned topic, showcasing their creativity, innovative thinking, and mathematical understanding. Throughout this phase, teams actively engaged in designing game mechanics, rules, and objectives that integrated the assigned mathematical concepts engagingly and interactively.

Finally, teams presented their newly invented games. Each allocated three minutes for their presentation. During this time, teams explained their games' rules, mechanics, and educational objectives, highlighting the mathematical concepts incorporated and potential learning outcomes. This presentation phase allowed teams to demonstrate their communication skills, creativity, and the application of mathematical principles in a game context.

Points were awarded based on the accuracy and creativity of the solutions developed during the problem-solving phase. Additionally, the newly invented games were evaluated based on their originality, educational value, and integration of mathematical concepts.

The organization of the fourth game provided a structured framework for teams to engage in mathematical problem-solving and unleash their creativity in designing new games. Participants could showcase their abilities in applying mathematical concepts to practical scenarios by combining logical thinking, mathematical skills, and innovative ideas. The game format encouraged autonomy, critical thinking, and the development of novel approaches to mathematical challenges.

2.8.5 Game 5: Technology

The fifth and final game of the STEMM project, focused on technology, was organized to encourage resilience, problem-solving, collaboration, and curiosity among the participating teams. The game revolved around the challenge of creating a compact closing mechanism for an A3-sized panel: "The Great in the Small."

Drawing inspiration from the mechanisms used by astronauts to open panels, teams were tasked with designing an efficient and practical closing mechanism [41].

The organization of the game involved several stages. Initially, teams were given 1 hour and 5 minutes for the folding process. During this time, teams collaboratively explored different folding techniques, shared ideas, and strategically experimented with various approaches to create the most effective closing mechanism. Cameras were on, allowing teams to communicate and observe each other's progress visually.

Following the folding phase, teams were granted three opportunities for constructive requests, during which they could seek clarification or specific instructions related to their design. This stage encouraged critical thinking as teams identified areas for improvement and sought expert guidance.

Finally, teams had 2 minutes to present their closing mechanism, highlighting its practicality, efficiency, and compatibility with the assigned A3-sized panel. The designs were evaluated based on the quality of constructive requests and the overall practicality of the proposed mechanism.

Throughout the game, participants utilized their resilience, problem-solving skills, collaboration, and curiosity to develop innovative closing mechanisms within the given timeframe. The organized structure allowed teams to approach the challenge creatively and ingeniously while fostering a spirit of friendly competition. By combining their knowledge of technology, problem-solving abilities, and teamwork, participants could showcase their skills in designing practical and efficient closing mechanisms for the A3-sized panel. While engaging in this interactive and hands-on challenge, participants not only developed a deeper understanding of technological concepts but also honed their abilities to solve complex problems, work effectively as a team, and think creatively in the realm of technology.

Soft skills played a vital role throughout the STEMM games project. Each session targeted specific soft skills and provided opportunities for students to develop and apply them in a practical context. By working in diverse teams with complementary skills and backgrounds, students had the chance to leverage their collective strengths, communicate effectively, and learn from one another. The interactive and competitive nature of the STEMM games required students to think critically, be creative, and adapt to dynamic situations. They had to strategize, plan their actions, and make quick decisions while working under time constraints. These challenges allowed students to refine their problem-solving abilities, enhance their resilience, and cultivate an open mindset.

The students participating in the STEMM project perceived the challenges as stimulating and enriching experiences. Through their feedback and reflections, it was evident that they embraced the opportunities presented by the project with enthusiasm and a sense of purpose. The challenges were seen as platforms for personal and collective growth, allowing them to apply their knowledge, skills, and creativity in practical and engaging ways.

The students expressed high motivation and engagement throughout the project, finding the challenges intellectually stimulating and rewarding. They appreciated the multidisciplinary nature of the games, which provided them with a broader perspective and allowed them to explore various fields within STEMM subjects.

The opportunity to work collaboratively in teams further enhanced their experience, as it fostered camaraderie, cooperation, and the sharing of diverse perspectives.

Moreover, the students recognized the value of the project in developing their soft skills. They acknowledged that the challenges not only tested their technical knowledge but also emphasized the importance of skills such as problem-solving, critical thinking, time management, creative thinking, and teamwork. By actively participating in the games, they acquired practical experience in applying these transferable skills, preparing them for future challenges in their academic and professional lives.

Overall, the students perceived the challenges as empowering and transformative, enabling them to grow both academically and personally. They expressed gratitude for the opportunity to engage in hands-on, experiential learning that went beyond traditional classroom settings. The challenges provided them a platform to push their boundaries, explore their capabilities, and develop a broader understanding of the STEMM fields, ultimately equipping them with valuable skills and perspectives for their future endeavors.

In conclusion, the STEMM games gave students a unique opportunity to develop soft skills while engaging playfully and interactively with STEMM subjects. The multidisciplinary nature of the project, combined with the competitive aspect, created a rich learning environment that nurtured creativity, collaboration, and critical thinking. By integrating game-based learning principles into EE, the project successfully demonstrated the effectiveness of this approach in enhancing students' soft skills and promoting holistic development. By leveraging the power of games, students were able to develop essential soft skills alongside their technical knowledge. The project showcased the effectiveness of this approach in fostering collaboration, problem-solving, critical thinking, and other transferable skills. The participating students' positive feedback and self-reported benefits indicate the value of incorporating STEMM games into engineering curricula to enhance the educational experience and prepare students for future career challenges.

2.9 The Importance of Laboratory Activities for Comprehensive Understanding: A Remote Laboratory Experience

Laboratory activities play a crucial role in EE as they provide students with hands-on experience and a deeper understanding of theoretical concepts. The practical application of knowledge in a controlled environment enhances students' problem-solving skills critical thinking abilities and fosters a deeper connection with the subject matter. However, traditional laboratory setups often face limitations regarding accessibility, resource allocation, and scheduling conflicts. To address these challenges and promote more active learning experiences, implementing remote laboratories has gained significant attention in recent years.

Recognizing the importance of hands-on learning, the Turin Polytechnic University in Tashkent (TTPU) in Uzbekistan identified the need to enhance its educational approach and incorporate more active teaching methodologies. In the next chapter, we will shift our focus to the perspective of teachers, and we will describe more in-depth the GYM: Grow You Methodology call (see Sect. 3.2.3) [5]. The TTPU is a private institution set up in the Spring of 2009 based on an agreement between the governmental entity in the automotive market, Uzavtosanoat SC, General Motors Corporation, and PoliTo. Both local and Italian professors jointly deliver TTPU lectures.

Implementing a remote laboratory into an Automatic Control course was one of the ideas awarded, which resulted in creating the Remote Experimentation for Automation Learning (REAL) platform [11]. The experimental activities on modeling and design control systems were performed in the Autumn of 2019 in the bachelor course at TTPU on a set of actual laboratory processes available in PoliTo through a suitable remote laboratory software/hardware platform.

Based on this first positive experience, the remote Automatic Control lab has been further extended in order to be also exploited in the course of Laboratory of Robust Identification and Control (LRIC), offered in Autumn 2020 to the students of the Master of Science in Mechatronic Engineering at PoliTo. The extension of the laboratory activities to Italian courses at PoliTo opened up new avenues for enhancing the educational experience of Italian students as well. By integrating the REAL platform into their curriculum, PoliTo aimed to provide their students with the same hands-on learning opportunities and practical insights that were being offered to their Uzbek counterparts. This extension of the laboratory activities to Italian courses not only broadened the reach and impact of the REAL platform but also facilitated students' independence in accessing laboratory activities.

The REAL remote laboratory implementation timeline, as depicted in Fig. 2.17, showcases the various stages of development, testing, and deployment. This timeline highlights the iterative design process, which involved the collaboration between educators, engineers, and students from TTPU and PoliTo.

The first edition of REAL (REAL 1.0) was made by a computer located at PoliTo in Turin, connected to a set of magnetic levitation systems. The remote

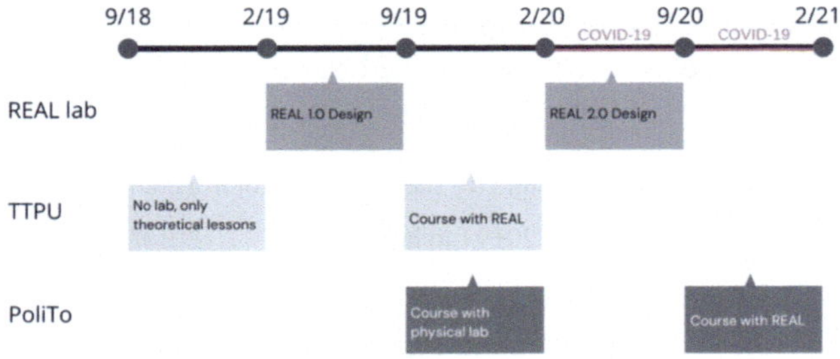

Fig. 2.17 REAL design implementation timeline

lab architecture consists of two parts: a client, used by the student, and a server, installed on a computer in the PoliTo lab. The students can study and simulate the Matlab/Simulink environment's control problem and then move to the real plant. Through a suitable Matlab library distributed to the students, they can connect to the computer located at the PoliTo lab, program the controller with their parameters, and test the system's performance. The students have direct control of the magnetic levitator, which can be switched on and off. A dedicated webcam pointed to the systems allows the student to visualize the experiment in real time. During the investigation, the student receives accurate measurements of the interest signals that can be plotted in real time or stored for a deferred analysis. In the second edition (REAL 2.0), the system has been modified and extended to handle the significantly more number of students attending the LRIC course. More precisely, a stand-alone software package based on Labview has been designed and made freely available to the students. Using such software, the students can directly and simultaneously connect to the REAL system to collect experimental data and/or implement their control algorithms without involving Matlab or any other proprietary software. Furthermore, the second edition of REAL also includes an online reservation system allowing the students to book the desired lab system for one or more specific time slots in the week.

Introducing the REAL remote laboratory has brought numerous benefits to students and international education. Firstly, the accessibility and flexibility of the remote laboratory allow students to engage in hands-on activities at any time, expanding learning opportunities beyond traditional classroom hours. This feature is precious for students facing scheduling constraints or limited access to physical laboratory facilities.

The performance indicators presented in Table 2.11 demonstrate the positive impact of REAL on the Automatic Control course at TTPU. The increased attendance rate, higher exam success rate, improved average marks, and a notable rise in the percentage of students achieving the highest marks reflect the effectiveness of the remote laboratory in enhancing student learning outcomes.

Similarly, Table 2.12 presents the performance indicators for the LRIC course at PoliTo. Despite the challenges posed by the COVID-19 pandemic, the remote laboratory maintained a high attendance rate, consistent exam success rate, improved average marks, and a sustained percentage of students achieving the highest marks.

The optimization of resources through the implementation of remote laboratories, such as REAL, not only enhances the educational experience for students but

Table 2.11 REAL performance indicators for the automatic control course at TTPU

Performance indicator	2018/2019 (no lab activity)	2019/2020 (with REAL)
Students	53	82
Attendance rate	8%	35%
Exam success rate	48%	58.5%
Average mark	22.4/30	24.7/30
% of highest mark (30/30)	6%	15%

Table 2.12 REAL performance indicators for the LRIC course at PoliTo

Performance indicator	2019/2020 (live lab activity)	2020/2021 (with REAL)
Students	80	125
Attendance rate	85%	78%
Exam success rate	77%	77%
Average mark	27.8/30	29.0/30
% of highest mark (30/30)	71%	72%

also supports educational institutions in developing countries. The collaboration between TTPU and PoliTo has successfully implemented a remote laboratory, demonstrating its effectiveness in promoting active learning, improving student engagement, and enhancing learning outcomes. The positive outcomes observed in both the Uzbek and Italian contexts highlight the potential of remote laboratories in offering inclusive and high-quality education.

In the next chapter, in particular in Sect. 3.2.4, we will shift our focus to the perspective of lecturers and explore the impact of the REAL remote laboratory on their teaching practices and professional development.

2.10 Bridging the Gap: Implementing Project-Based Learning to Teach Engineering Concepts to Architecture Students

PoliTo faces a unique challenge in supporting architectural courses with engineering subjects. While the course content may overlap, architecture students' backgrounds and skill sets differ significantly from those of engineering students. This presents an excellent opportunity to explore the implementation of problem-based learning (PBL) methodology in the Fundamentals of Structural Analysis (FSA) course, catering to a diverse international audience of architectural students with varying levels of technical expertise. This section delves into the application of PBL in the FSA course, aiming to bridge the gap between architectural and EE while nurturing critical problem-solving skills and interdisciplinary collaboration.

PBL is an educational approach that places the problem itself at the core of the learning process, as opposed to the traditional didactic teaching based on the transmission of theoretical concepts. In the context of EE, PBL has been widely adopted due to its promotion of active student engagement, teamwork, and the acquisition of practical skills.

PBL involves the active participation of students in solving real or simulated problems, stimulating self-directed learning, research, and critical reflection. Students are presented with complex situations and technical challenges that require applying theoretical knowledge to develop concrete solutions. This approach aligns well with the demands of the professional world, where engineers must tackle intricate problems and find innovative solutions.

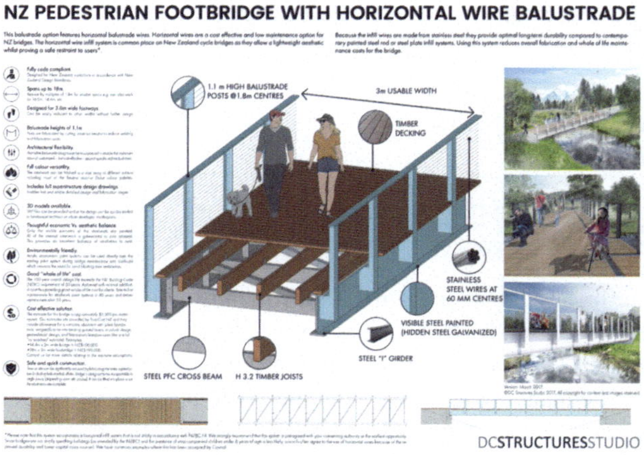

Fig. 2.18 New Zealand standard footbridge design (elementary 18-m pedestrian bridge)

FSA is a mandatory third-year course in the architecture bachelor program at PoliTo. The FSA course challenges architectural students due to its technical and engineering content, especially in understanding theoretical concepts and manual and numerical solution procedures. The FSA course at PoliTo has traditionally followed a lecture-based model, with a single instructor using slides and a whiteboard to deliver the content. However, this methodology encountered several difficulties, especially in engaging students and promoting the transition from theory to qualitative analysis, which is particularly challenging for architecture students.

The redesign of the FSA course to adopt PBL as a teaching methodology was inspired by several studies on its effectiveness in technical and engineering contexts. In the case of architecture, this change required a degree of creativity and adaptability, but it proved highly beneficial in addressing student learning difficulties and improving their engagement.

The chosen project for the FSA course is the design and structural analysis of a small pedestrian bridge inspired by a real case study (Fig. 2.18). This project offers a blend of theoretical and practical aspects, encouraging students to consider a wide range of skills and knowledge acquired during their academic journey. For example, it aims to foster creativity, problem-solving, and interdisciplinary thinking by allowing students to select their bridge's location and building materials based on their cultural backgrounds and experiences.

To support the students throughout the PBL implementation, a team of tutors, selected from civil engineering master students, were assigned to assist and guide the teams during their project work. These tutors are crucial in facilitating learning and ensuring equal task distribution within the teams. Their guidance and support are essential to help students overcome initial disorientation and maximize their conceptual understanding of structural analysis.

The course calendar was structured into three main components: lectures, tutoring sessions, and presentations. The lectures covered theoretical concepts, computational exercises, and essential content related to the project. The tutoring sessions allowed students to discuss their projects with the assigned tutors, ensuring personalized guidance and feedback. The presentations offered students the opportunity to showcase their progress and ideas to the class and external experts, such as the DC Structures Studio of Cambridge (NZ), which was involved in the evaluation process.

The application of PBL in the FSA course has yielded several significant advantages:

- **Active student engagement** PBL has empowered students to actively participate in their learning, encouraging them to investigate, explore, and solve problems independently. This active involvement has increased student interest in the course and improved their motivation to understand theoretical concepts through practical application.
- **Development of transversal skills** PBL promotes teamwork and peer-to-peer learning. Working in teams of 3–4 students encourages collaboration and helps develop essential soft skills, such as communication, organization, time management, and leadership, which are crucial for future architecture professionals, as they often collaborate with colleagues from diverse disciplines.
- **Critical reflection** PBL fosters critical reflection on the outcomes of the learning process. Students are encouraged to evaluate their knowledge, skills, and problem-solving approach. By encountering authentic challenges in designing a pedestrian bridge, students learn to critically analyze structural aspects, consider various materials and locations, and make informed decisions based on research and analysis. This reflection helps them develop awareness about their learning and autonomy in improving their competencies.
- **Active and meaningful learning** PBL has enabled students to connect previously acquired knowledge with the pedestrian bridge problem. This connection has made learning more meaningful as students comprehend how to apply theory to real-world situations. It also increases students' motivation and interest in the subject matter, leading to improved attendance and participation during lectures.

The results obtained from the initial implementation of PBL in the FSA course have been generally positive, as presented at the EDUCON 2020 conference [8] and in a long paper [10]. The participation of students has increased significantly compared to previous academic years, indicating higher engagement and interest in the subject matter. The interaction between the lecturer and students has also improved, with increased communication through emails and active learning activities implemented during lectures.

Students have embraced the freedom to choose bridge locations and materials, boosting their interest and motivation to explore design solutions aligned with their cultural backgrounds. Using tutors has been instrumental in guiding students and ensuring equitable task distribution within teams, further supporting students' learning outcomes.

Final assessment data showed a significant improvement in the number of students who passed the course exam (Table 2.13). Compared to the previous

Table 2.13 Students' distribution at the assessment and their average score

Exam session	Academic year 2018/2019			Academic year 2019/2020			
	# students	Test results	Final grade	# students	Test results	Project results	Final grade
Winter	5 (11%)	21,4 /30	21,4 /30	41 (66%)	14,82 /18	10,29 /12	25,26 /30
Summer + Autumn	33 (73%)	23,2 /30	23,2 /30	11 (18%)	14,18 /18	8,55 /12	22,78 /30
None	7 (16%)			10 (16%)			
Total	45			62			

academic year, there was an increase both in the number of students who passed the exam and in the average grades obtained.

The application of PBL in the FSA course has been demonstrated to be an effective educational methodology in teaching engineering concepts to nonengineering students, particularly within the context of architecture. Students appreciated the opportunity to address a real and relevant project, enabling them to apply theoretical knowledge practically. Active involvement and autonomy in developing solutions have led to an improved learning process and the acquisition of essential transversal skills. Students have shown increased interest in the course and heightened awareness of their own learning.

However, there were some challenges, such as monitoring the progress of individual team projects and adapting the methodology to accommodate varying student competencies. Language difficulties for non-native speakers also required additional resources to ensure proper comprehension of technical concepts. Some of these aspects will be discussed in the next chapter that considers the lecturer's perspective (see Sect. 3.2.4)

Overall, the application of PBL in the FSA course has proven successful and serves as a solid foundation for further development and enhancement of EE. The methodology has been shown to prepare students to become competent and conscientious future professionals capable of creatively and innovatively addressing real-world challenges.

In summary, this chapter has explored the educational pathways of students in STEM, highlighting the importance of preliminary preparation, the influence of higher education, and the crucial role of gender diversity in engineering. Students, as the protagonists of their educational journey, face challenges and opportunities that shape their skills and future perspectives. Through initiatives such as the TIL and the CIAO!, PoliTo has shown a significant commitment to supporting students, ensuring a solid and inclusive educational path. The STEMM project, with its playful and interdisciplinary activities, has further emphasized the importance of soft skills, making learning more dynamic and engaging. Finally, the Talenti project and the reverse inclusion approach demonstrate how targeted support for talented students can generate widespread benefits for the entire student community. The practical implications of these initiatives are profound: They offer students enhanced learning experiences through hands-on and PBL, promote equitable access to educational resources, and equip students with critical skills for their future careers. Looking ahead, it is essential to continue innovating and adapting educational methodologies to support students in an ever-evolving world, cultivating the next generation of capable and diverse engineers.

References

1. Ballatore MG, Tabacco A (2021) Daybyday: a continuous engagement for better learning in a first-year mathematical course. In: Proceedings—SEFI 49th Annual Conference. Sefi, pp 1347–1353. ISBN 978-2-873520-23-6
2. Ballatore MG, Tabacco A (2024) Enhancing soft skills in applied STEM fields through games. In: 2024 IEEE Global Engineering Education Conference (EDUCON), pp 1–5. https://doi.org/10.1109/EDUCON60312.2024.10578857
3. Ballatore MG, Montanaro L, Tabacco A (2018) Til: an innovative tool for the recruitment of bachelor engineering students in Italy. Int Educ Res J 4(2):79–84. Retrieved from http://ierj.in/journal/index.php/ierj/article/view/1504
4. Ballatore MG, Montanaro L, Tabacco A (2019) Empowering talented students: an Italian experience of an enriched curriculum in engineering. Int J Eng Pedagogy 9(3):56–75. https://doi.org/10.3991/ijep.v9i3.10140
5. Ballatore MG, Stievano IS, Tabacco AM (2019) Teach-gym: grow your methodologies. In: Innovations, Technologies and Research in Education. Linda Daniela—University of Latvia, Latvia, pp 180–191. https://doi.org/10.22364/atee.2019.itre. ISBN 978-99-34184-82-6
6. Ballatore MG, Montanaro L, Tabacco A (2019) Inclusion: a new reverse perspective. In: Realising Ambitions: Proceedings of the 6th Annual Symposium of the United Kingdom & Ireland Engineering Education Research Network.
7. Ballatore MG, Barman L, De Borger J, Ehlermann J, Fryers R, Kelly K, Misiewicz J, Naimi-Akbar I, Tabacco A (2019) Increasing gender diversity in STEM. In: Proceeding TEEM'19 Proceedings of the Seventh International Conference on Technological Ecosystems for Enhancing Multiculturality. ACM, pp 216–222. https://doi.org/10.1145/3362789.3362832. ISBN 978-14-50371-91-9
8. Ballatore MG, Barpi F, Tabacco A (2020) Work-in-progress: pedestrian bridge application in a fundamentals of structural analysis course inside an architecture bachelor program. In: IEEE Global Engineering Education Conference, EDUCON. IEEE Computer Society, pp 41–45. https://doi.org/10.1109/EDUCON45650.2020.9125140. ISBN 978-1-728109-30-5
9. Ballatore MG, Duffy G, Sorby S, Tabacco A (2020) SAperI: approaching gender gap using Spatial Ability training week in high-school context. In: Eighth International Conference on Technological Ecosystems for Enhancing Multiculturality (TEEM'20). ACM, pp 142–148. https://doi.org/10.1145/3434780.3436577. ISBN 978-14-50388-50-4. Retrieved from https://dl.acm.org/doi/10.1145/3434780.3436577
10. Ballatore MG, Barpi F, Crocker D, Tabacco A (2021) Pedestrian bridge application in a fundamentals of structural analysis course inside an architecture bachelor program. Int J Eng Pedadogy 11(5):4–18. International Association of Online Engineering (IAOE). https://doi.org/10.3991/ijep.v11i5.20151. Retrieved from https://doi.org/10.3991/ijep.v11i5.20151
11. Ballatore MG, Razza V, Regruto Tomalino D, Stievano IS, Tabacco AM (2022) The educational benefit of a remote automatic control laboratory. A win-win collaboration between Asia and Europe. Educ Knowl Soc 23. Universidad de Salamanca. https://doi.org/10.14201/eks.28495. Retrieved from https://revistas.usal.es/index.php/eks/article/view/28495
12. Ballatore MG, Damonte L, Tabacco AM (2023) Maths games without frontiers. In: To Be or Not to Be a Great Educator, 2022. Proceedings of ATEE Annual Conference. University of Latvia—Latvijas Universitāte, pp 642–651. https://doi.org/10.22364/atee.2022.43. ISBN 978-99-34360-19-0

13. Bawa P (2020) Game on! Int J Game-Based Learn 10(3):16-46. IGI Global
14. Bennett G (1973) Differential Aptitude Tests: Forms S and T. Psychological Corporation, New York
15. Burton LJ, Fogarty GJ (2003) The factor structure of visual imagery and spatial abilities. Intelligence 31(3):289–318. Elsevier BV. https://doi.org/10.1016/S0160-2896(03)00061-8
16. Carr M, Davis H (2001) Gender differences in arithmetic strategy use: a function of skill and preference. Contemp Educ Psychol 26(3):330–347. https://doi.org/10.1006/ceps.2000.1059
17. Carroll JB (1993) Human Cognitive Abilities: A Survey of Factor-Analytic Studies. Cambridge University Press, Cambridge
18. Casey MB, Nuttall R, Pezaris E, Benbow CP (1995) The influence of spatial ability on gender differences in mathematics college entrance test scores across diverse samples. Dev Psychol 31(4):697–705. American Psychological Association (APA)
19. Cohen CA, Hegarty M (2012). Inferring cross sections of 3D objects: a new spatial thinking test. Learn Individ Differ 22(6):868–874. Elsevier BV. https://doi.org/10.1016/j.lindif.2012.05.007
20. Coluccia E, Louse G (2004) Gender differences in spatial orientation: a review. J Environ Psychol 24(3):329–340. https://doi.org/10.1016/j.jenvp.2004.08.006
21. Crocco F, Offenholley K, Hernandez C (2016). A proof-of-concept study of game-based learning in higher education. Simul Gaming 47(4):403–422. SAGE Publications. https://doi.org/10.1037/0033-2909.86.5.889
22. Delgado A, Prieto G (2004) Cognitive mediators and sex-related differences in mathematics. Intelligence 32:25–32. https://doi.org/10.1016/S0160-2896(03)00061-8
23. Freedman RJ, Rovegno L (1981) Ocular dominance, cognitive strategy, and sex differences in spatial ability. Perceptual Motor Skills 52(2):651–654. https://doi.org/10.2466/pms.1981.52.2.651
24. Ghiselli EE (1973) The validity of aptitude tests in personnel selection. Personnel Psychol 26(4):461–477. https://doi.org/10.1111/j.1744-6570.1973.tb01150.x
25. Halpern DF, Benbow CP, Geary DC, Gur RC, Hyde JS, Gernsbacher MA (2007) The science of sex differences in science and mathematics. Psychol Sci Public Inter 8(1):1–51. https://doi.org/10.1111/j.1529-1006.2007.00032.x
26. Hawes Z, Ansari D (2020) What explains the relationship between spatial and mathematical skills? A review of evidence from brain and behavior. Psychon Bull Rev 27(3):465–482. Springer Science and Business Media LLC. https://doi.org/10.1007/s10648-015-9337-9
27. Kimura D (1999) Sex and Cognition. MIT Press, Cambridge
28. Lappan G (1981) Middle Grades Mathematics Project: Spatial Visualization Test. MSU Mathematics Dept, Middle Grades Mathematics Project
29. Levine SC, Huttenlocher J, Taylor A, Langrock A (1999) Early sex differences in spatial skill. Dev Psychol 35(4):940–949. American Psychological Association (APA)
30. Levine SC, Vasilyeva M, Lourenco SF, Newcombe NS, Huttenlocher J (2005) Socioeconomic status modifies the sex difference in spatial skill. Psychol Sci 16(11):841–845. https://doi.org/10.1111/j.1467-9280.2005.01623.x
31. Liben LS, Colbeck SL (1980) Sex differences in performance on Piagetian spatial tasks: differences in competence or performance? Child Develop 51:594–597
32. Linn MC, Petersen AC (1985) Emergence and characterization of sex differences in spatial ability: a meta-analysis. Child Develop 56(6):1479–1498. Wiley, Society for Research in Child Development. Retrieved from http://www.jstor.org/stable/1130467
33. Lucisano P, du Mérac ÉR (2019) Soft skills self-evaluation questionnaire (3SQ), caratteristiche e proprietà psichometriche. In: Lucisano P, Notti A (eds) Convegno Internazionale SIRD: Training Actions and Evaluation Processes. Pensa Multimedia, Lecce, pp 609–622
34. Masters MS, Sanders B (1993) Is the gender difference in mental rotation disappearing? Behavior Genet 23(4):337–341. https://doi.org/10.1007/BF01067434
35. McGee MG (1979) Human spatial abilities: psychometric studies and environmental, genetic, hormonal, and neurological influences. Psychol Bull 86(5):889–918. https://doi.org/10.1037/0033-2909.86.5.889

36. Mercuri G (2019) La sindrome di Hermione: una spiegazione sul perché le ragazze vincono a scuola e perdono al lavoro. Corriere della Sera
37. Meyer M, Koehler MS (1990) Internal influences on gender differences in mathematics. In: Fennema E, Carpenter LG (eds) Mathematics and Gender. Teachers College Press, New York, pp 60–95
38. Mix KS, Cheng Y-L (2012) The relation between space and math. In: Advances in Child Development and Behavior, vol 42. Elsevier, Amsterdam, pp 197–243
39. Moè A, Meneghetti C, Cadinu M (2009) Women and mental rotation: incremental theory and spatial strategy use enhance performance. Personal Individ Differ 46(2):187–191. https://doi.org/10.1016/j.paid.2008.09.030
40. Morando P (2009) Learning and playing mathematics. In: EDULEARN09 Proceedings. IATED, pp 3038–3043. ISBN 978-84-61298-01-3
41. Museo delle Scienze, MUSE (2018) Il messaggio di Paolo Nespoli al MUSE dalla stazione ISS. YouTube. Retrieved from https://www.youtube.com/watch?v=vu9cFGwD2Ps
42. Newcombe N (2013) Seeing relationships: using spatial thinking to teach science. Am Educ 37:26–40
43. Nordvik H, Amponsah B (1998) Gender differences in spatial abilities and spatial activity among university students in an egalitarian educational system. Sex Roles 38(11/12):1009–1023. https://doi.org/10.1023/A:1018878610405
44. Plass JL, Homer BD, Kinzer CK (2015) Foundations of game-based learning. Educ Psychol 50(4):258–283. Informa UK Limited. https://doi.org/10.1002/j.2168-9830.2011.tb00003.x
45. Sorby S (2005) Assessment of a "New and Improved" course for the development of 3-D spatial skills. Eng Design Graph J 69(3):6–13
46. Sorby SA, Baartmans BJ (2000) The development and assessment of a course for enhancing the 3-D spatial visualization skills of first year engineering students. J Eng Educ 89(3):301–307. Wiley. https://doi.org/10.1002/j.2168-9830.2000.tb00529.x
47. Voyer D, Voyer S, Bryden MP (1995) Magnitude of sex differences in spatial abilities: a meta-analysis and consideration of critical variables. Psychol Bull 117(2):250–270. https://doi.org/10.1037/0033-2909.117.2.250
48. Wai J, Lubinski D, Benbow CP (2009) Spatial ability for STEM domains: aligning over 50 years of cumulative psychological knowledge solidifies its importance. J Educ Psychol 101(4):817–835
49. Yoon SY (2011) Psychometric properties of the Revised Purdue Spatial Visualization Tests: Visualization of Rotations (The Revised PSVT). (Doctoral dissertation). Purdue University

Chapter 3
Enhancing Engineering Education from the Lecturer's Perspective

> *Three bricklayers are working hard to build a building. A passerby, curious, asked the three of them, 'What are you doing?' The first bricklayer, tired, annoyed, and sweaty, replied, 'I am building a wall.' The second bricklayer, sighing, answered, 'I am building a cathedral.' The third bricklayer, despite the heat, fatigue, and scorching sun, replied with a smile, 'I am building the house of God, where He will be glorified for centuries to come.'*
>
> — *Anonymous, Parable*

Abstract The chapter is structured to encompass a spectrum of topics, each contributing to a holistic understanding of the lecturer's role in shaping effective EE. It embarks on a journey through training opportunities that fortify both primary and secondary education, elucidating how educators at all levels play a collaborative role in nurturing engineering foundations. The journey then transitions to professional development for university-level educators, unveiling the tools and platforms designed to empower their pedagogical approaches. The exploration further extends to the integration of open educational practices and methodologies that optimize course delivery, ultimately culminating in a global perspective that highlights the cross-cultural exchange and adaptability necessary for the modern engineering educator.

As the chapter unfolds, it illuminates the interconnectedness of these components, reinforcing the understanding that effective EE is a synergy of several vital elements. Through this exploration, the chapter seeks not only to shed light on the challenges and opportunities faced by educators but also to celebrate the collective efforts that contribute to the advancement of EE as a whole.

Through these pages, we invite readers to delve into the intricacies of EE from a vantage point that is both pivotal and transformative—that of the dedicated educator.

In the dynamic field of engineering education (EE), the pivotal role of university lecturers cannot be overstated. These educators stand at the forefront of knowledge

M. G. Ballatore, *Empower Education to Foster Engineering*,
PoliTO Springer Series, https://doi.org/10.1007/978-3-031-75895-9_3

dissemination, inspiring the next generation of engineers and shaping the trajectory of the field. As the driving force behind the learning experience, their commitment to innovation and professional development significantly influences the quality and effectiveness of EE.

As the chapter unfolds, it illuminates the interconnectedness of these components, reinforcing the understanding that effective EE is a synergy of several vital elements. Through this exploration, the chapter seeks not only to shed light on the challenges and opportunities faced by educators but also to celebrate the collective efforts that contribute to the advancement of EE as a whole.

In this pursuit, we embark on an exploration of the myriad ways in which EE can be nurtured, refined, and innovated from the lecturer's perspective. By delving into the training opportunities, instructional methodologies, and global influences that shape this multifaceted field, we aim to contribute to a comprehensive understanding of how educators play a pivotal role in molding the engineers of tomorrow. Join us on this journey as we navigate the realms of training, methodologies, and international perspectives, uncovering the threads that weave the tapestry of excellence in EE.

3.1 Training Opportunities and Professional Development

3.1.1 Reinforcing Engineering Prerequisites in K-12 Education

In the landscape of modern education, the significance of engineering as a cornerstone of innovation and societal advancement is widely acknowledged. An essential facet of fostering a skilled and informed cohort of future engineers lies in the systematic integration of EE at the primary and secondary levels. This subsection delves into the pivotal role of bolstering engineering prerequisites during the formative years of K-12 education and examines the empirical insights offered by recent research.

Within the context of K-12 education, the introduction of engineering concepts has emerged as a pivotal avenue for nurturing future engineers and fostering a culture of innovation. Notably, the studies conducted by Genalo [13] and Katehi et al. [22] emphasize the significance of early exposure to EE. Genalo advocates for the training of educators to craft engineering-infused learning experiences, underscoring its centrality to curricular enhancement [13]. Katehi, Pearson, and Feder's research expounds on the multifaceted benefits of integrating engineering into K-12 classrooms, ranging from improved student learning outcomes in science and Mathematics to heightened awareness of engineering as a discipline and career path [22].

Numerous scholarly inquiries have further substantiated the value of early EE. The work of Olds [26] introduces a K-12 engineering outreach project with a twofold aim: benefiting both the recipients of instructional materials and the project developers. This duality underscores the reciprocal advantages of engineering exposure at a young age, fostering a symbiotic relationship between educational advancement and curriculum enhancement. Dornian, Moshirpour, and Behjat's systematic literature review [11] extends the discourse by underscoring the preparatory role of K-12 digital education in equipping future engineering students with enhanced learning outcomes in the twenty-first century.

The studies conducted by Ghalia et al. [14], Guzey et al. [15], Porter et al. [29], and Singer et al. [31] converge on a crucial theme: the transformative potential of targeted professional development for educators. These investigations affirm that equipping teachers with the tools and pedagogical strategies to integrate engineering into their classrooms is a pivotal step towards strengthening engineering prerequisites. Ghalia and Guzey highlight the successful implementation of engineering design lessons by teachers who participated in professional development programs [14]. Porter's research underscores the importance of training and guidance to seamlessly embed engineering concepts [29], while Singer's findings reveal the positive impact of professional development on high school STEM teachers' ability to enact design-based pedagogical practices associated with engineering curricula [31].

John Hattie's research on visible learning adds another layer of insight into the effectiveness of integrating engineering concepts into K-12 education. Hattie's extensive meta-analyses identify the key influences on student achievement, emphasizing the critical role of teachers in the learning process [16]. Visible learning posits that when teachers see learning through the eyes of their students and help students become their own teachers, the impact on learning is profound. Applying Hattie's principles to EE suggests that professional development programs should not only equip teachers with technical skills but also enhance their ability to reflect on and adapt their teaching strategies based on student feedback and performance data. This reflective practice aligns with the principles of visible learning, promoting a deeper understanding of engineering concepts and fostering a more effective learning environment.

3.1.1.1 The Matabì Project: Bridging Gender Gaps in Mathematics

The Matabì training course represents a pioneering initiative in professional development tailored specifically for primary school educators. Its meticulous design aims to enhance Mathematics education and address gender disparities in the classroom. Developed collaboratively by Fondazione Agnelli and Politecnico di Torino (PoliTo), with generous support from Exor and The LEGO Foundation, the course is part of the Matabì projects.

At its core, the Matabì project targets the reduction of the gender gap in math and science by integrating manipulative elements into traditional mathematical lessons for third- and fourth-grade elementary students. Embracing constructivist and inquiry-based educational philosophies, this pedagogical tool fosters hands-on, interactive learning experiences that cultivate spatial thinking and problem-solving

Fig. 3.1 Matabì kit

skills among students. It encourages active construction of mathematical knowledge through exploration, experimentation, and collaboration.

Central to the Matabì approach is its meticulously curated student kit, illustrated in Fig. 3.1. Comprising SixBrick elements [18], long 2×6 bricks in red and light green, two square bricks in corresponding colors, and thinner dark blue bricks, this kit is designed to bridge the abstract nature of mathematical concepts with students' concrete understanding. Drawing on insights from cognitive psychology and educational neuroscience, Matabì leverages spatial abilities to facilitate learning, enabling students to explore geometric concepts, visualize fractions, and solve word problems through physical manipulation.

Designed for seamless integration into existing curricula, Matabì empowers teachers to provide students with enriching hands-on experiences while catering to diverse learning needs through differentiated instruction. This approach not only reinforces mathematical understanding but also fosters peer-to-peer interactions, creating a conducive and empowering learning environment without necessitating changes to the curriculum contents.

Central to the Matabì project is the teacher training course. This is an intricately structured journey designed to empower primary school teachers with the skills and strategies required for fostering equitable Mathematics education, transcending gender-based disparities. Now we delve into the granular details of the training's content and structure, revealing its transformative pedagogical framework.

The course unfolds over a comprehensive 10-hour trajectory, melding both face-to-face and online formats to optimize accessibility and flexibility for educators. This curriculum encompasses five pivotal modules, each dedicated to specific facets of pedagogical refinement:

Module 1: Understanding Gender Stereotypes in Mathematics The journey commences with an exploration of gender stereotypes' impact on Mathematics education. Educators engage in illuminating discussions, dismantling misconceptions and acknowledging the societal influences shaping performance disparities. This module lays the foundation for addressing implicit biases, fostering a pedagogical landscape grounded in inclusivity.

Module 2: The Significance of Spatial Skills in Mathematics Spatial skills take center stage in the second module, as teachers embark on a voyage into the world of spatial thinking. Hands-on exercises, curated akin to students' experiences, beckon educators to refine their spatial abilities. This introspective journey fosters empathy and a deeper understanding of spatial challenges, laying the groundwork for enhanced teaching methodologies.

Module 3: Introducing the Matabì Kit The third module introduces the Matabì kit, an innovative tool bridging spatial exploration and Mathematics instruction. Educators familiarize themselves with the kit's components, unraveling its potential to infuse classrooms with experiential learning. This hands-on engagement primes teachers to weave spatial exploration into their pedagogical repertoire.

Module 4: Strategies for Teaching with Matabì Module four delves into pedagogical strategies that harmonize the Matabì kit's potential with instructional efficacy. Educators are immersed in best practices, understanding how to seamlessly integrate spatial exploration into existing curricula. This module unfolds strategies tailored to each grade, empowering teachers to catalyze transformative learning experiences.

Module 5: Personal Skill Development and Empowerment The final module circles back to personal skill development. Here, educators undergo training akin to students, grappling with spatial challenges and enhancing their own spatial skills. This introspective journey fortifies teachers' spatial acumen, instilling confidence and a nuanced perspective that transcends traditional pedagogy.

Workshops in class The culmination of the training is manifest in workshops and classroom implementation. Educators embark on a series of workshops, each meticulously curated to empower teachers in distinct aspects of spatial mathematics:

1. Angles or triangles: A tutor-led workshop that boosts educators' confidence in using Matabì for geometric concepts.
2. Measurements and equivalences: Teachers take the lead in this workshop, concentrating on measurements and equivalences, harnessing the Matabì kit's power.
3. Spatial mathematics with isometries: A teacher-led exploration delving into symmetry, translations, and rotations through hands-on brick activities.

Additionally, teachers create and execute an additional workshop, shaping personalized learning experiences for their students, echoing the core principle of Matabì's transformative journey.

The Matabì training course harmonizes face-to-face and online modalities, ensuring a flexible and comprehensive learning experience. The initial pilot phase involved 15 teachers and prioritized direct interaction, fostering immediate feedback and dynamic idea exchange. Transitioning to an online format in subsequent editions, with 69 teachers, introduced flexibility and accessibility, making instructional materials, video lectures, and interactive sessions available via a digital platform. The online format allows teachers to engage at their own pace, supplemented by virtual tutoring meetings and emails for ongoing support and engagement.

Central to the Matabì teacher training is a rich repository of supplementary didactical materials that empowers educators with a diverse toolkit for pedagogical

(a) (b) (c) (d)

Fig. 3.2 Variety of materials available. (**a**) Workbook. (**b**) Didactic cards. (**c**) Video clip. (**d**) Padlet

transformation. These materials extend beyond the conventional lecture-centric approach, fostering an immersive and interactive learning experience for teachers.

One cornerstone of this arsenal is the bespoke "workbook" meticulously curated to hone teachers' spatial skills (Fig. 3.2a). This compendium of exercises orchestrates a deliberate journey towards enhancing spatial thinking, acknowledging the educators as learners in their own right. Through these exercises, teachers navigate spatial challenges akin to those encountered by their students, fostering empathy, deeper understanding, and personal growth.

Complementing the workbook are the "didactic cards," strategically tailored to suit the distinct learning trajectories of third- and fourth-grade students (Fig. 3.2b). These cards orchestrate engaging activities that intertwine with the workshop sessions, offering teachers a dynamic toolkit to infuse spatial ability development within their classrooms. This interactive pedagogical resource empowers teachers to orchestrate hands-on spatial explorations, equipping students with cognitive skills that transcend the confines of traditional learning paradigms.

The Matabì training course is an immersive odyssey that empowers educators with the knowledge, skills, and transformative strategies required to reshape the contours of Mathematics education. By unraveling nuanced modules, fostering personal growth, and nurturing hands-on implementation, Matabì emerges as a beacon of innovation, poised to ignite transformative learning environments grounded in gender equity and spatial acumen. Acknowledging the multifaceted nature of modern education, the Matabì project strategically integrates multimedia elements into its training arsenal. Video clips, replete with comprehensive explanations, elucidate the nuances of workshops and activities involving the distinctive LEGO Duplo bricks (Fig. 3.2c). These visual aids serve as navigational beacons, guiding teachers through the intricacies of pedagogical execution.

Furthermore, the project leverages the dynamic capabilities of the digital world with the creation of a dedicated Padlet (Fig. 3.2d). This virtual platform nurtures collaboration and knowledge exchange among educators, facilitating the sharing of materials, insights, and innovative teaching strategies. The Padlet serves as an intellectual crossroads, catalyzing a dynamic community of educators vested in reshaping Mathematics pedagogy.

The empirical outcomes of the Matabì training course reverberate with transformation, both quantifiable and qualitative. Notably, the project's efficacy becomes palpable through observed improvements in teachers' spatial abilities despite their nonvoluntary involvement in the training program. The baseline assessment using the Revised PSVT:R [36] involved 76 teachers, who achieved an average score of 10.09 (SD=6.18), aligning with existing literature that locates the education community as the poorest one on spatial ability [35]. The end-line assessment included 52 teachers, with an increased average score of 12.94 (SD = 5.88). Notably, 47 teachers completed both the pre- and posttests, enabling a direct comparison of their performance. The two-sample t-test with equal variances revealed a significant improvement in spatial ability, with a mean increase of 2.57 ($t(92) = 2.14$, $p = 0.035$) in the number of correct answers in the posttest, corresponding to 25% increase in the score.

The observed improvement in spatial ability highlights the effectiveness of the Matabì project's interventions and training programs in enhancing teachers' spatial thinking skills. This indicates the project's potential to positively impact teachers' professional development and subsequently improve students' learning outcomes.

Moreover, our sample composition, which includes teachers who may not have been initially motivated to participate, strengthens the generalizability of the findings, providing a realistic portrayal of the broader teaching community. Additionally, comparing the baseline results with existing literature on spatial ability in STEM domains further reinforces the significance of spatial thinking skills in Mathematics education.

Furthermore, examining two pilot editions, one conducted face-to-face with deeper spatial learning and the other conducted online, revealed noteworthy differences. The face-to-face pilot exhibited a baseline mean of 9.08 and a posttest mean of 13.02, while the online pilot showed a baseline mean of 10.28 and a posttest mean of 12.88. Although the analysis is limited due to the difference in the number of teachers between the editions, it is evident that face-to-face training yielded better results. This advantage can be attributed to the facilitation of physical presence, which fosters collaboration, interaction, and longer-duration engagement for personal improvement.

This study, presented at the TEEM conference 2023 [7], focused solely on teachers' spatial ability and did not explore other potential outcomes or factors that may influence the effectiveness of the Matabì project. Future research should aim for a more comprehensive understanding of the impact.

Overall, our findings suggest that the Matabì project effectively enhances teachers' spatial ability, contributing to their professional development and potentially benefiting students' learning outcomes. These positive results underscore the potential of incorporating playful learning and LEGO-based activities in Mathematics education to foster spatial thinking and promote a deeper understanding of geometric and spatial concepts. The Matabì training course is an immersive odyssey that empowers educators with the knowledge, skills, and transformative strategies required to reshape the contours of Mathematics education. By unraveling nuanced modules, fostering personal growth, and nurturing hands-on implementation, Matabì emerges as a beacon of innovation, poised to ignite transformative learning environments grounded in gender equity and spatial acumen.

3.1.1.2 The SMaILE Project: Fostering AI Literacy in Early Education

In addition to the Matabì training course, another exemplary initiative that underscores the transformative potential of teacher training is the course inside the SMaILE project (Simple Methods for Artificial Intelligence Learning and Education). SMaILE is designed to pioneer game-based tools for teaching artificial intelligence (AI) concepts through experiential learning approach.

The SMaILE project is a research endeavor led by PoliTo in collaboration with the University of Torino and Royal Holloway University of London. This project emerged as the winner in the "Education and Training" category of the "Artificial Intelligence, Human, and Society" 2021 call, which the Compagnia di San Paolo Foundation promoted.

AI is rapidly becoming an indispensable component of our daily lives, permeating various societal domains. Its pervasive presence necessitates a proactive approach to incorporating AI education into the academic curriculum, considering its potential impact on future engineering professionals. Anticipating a surge in AI reliance in the form of autonomous driving mechanisms, robotic systems, and personalized medicine, it is imperative to equip students with a comprehensive understanding of AI principles and aptitudes.

Italian primary and secondary school curricula exhibit a striking need for more digital competencies encompassing AI-related knowledge and skills. Consequently, a pressing need arises to introduce AI as a discipline of study, accompanied by broader informatics competencies, and leverage it as a pedagogical tool to enhance the quality and efficacy of education. This critical exigency calls for integrating novel methodologies to facilitate the acquisition of digital literacy and AI awareness among young students while simultaneously optimizing the learning experience through the proficient utilization of AI.

The SMaILE project emerges as a pivotal intervention in this context, aiming to facilitate the effective utilization of AI tools while fostering an in-depth understanding of AI principles, codes, characteristics, and applications.

To accomplish its overarching objective, SMaILE harnesses the principles of Game Theory and employs gamification techniques as educational instruments. This strategic utilization aims to maximize student engagement, enabling profound internalization of knowledge and the practical acquisition of AI application skills.

The main objectives of the project can be summarized as follows:

- Facilitate the comprehension of AI principles and techniques among young learners
- Promote the effective and responsible utilization of AI tools across diverse disciplines and applications
- Formulate novel learning trajectories in AI education, addressing the prevailing curricular gaps
- Contribute to positioning the city of Turin as a pioneering hub and educational laboratory for AI instruction
- Disseminate the research project's outcomes to a broad audience, encompassing both national and international realms

The SMaILE project endeavors to bridge the existing gaps in AI education, empowering students with digital competencies and nurturing a future generation of engineers proficient in leveraging AI technology responsibly and knowledgeably.

Given the inherently multidisciplinary nature of AI, the SMAILE project adopts a comprehensive approach by engaging a team of mathematicians, computer scientists, engineers, and psychologists. This diverse team brings extensive research and teaching expertise in the field of AI, representing academic institutions, cultural associations, schools, and leading companies in the gaming and innovation sectors. The project encompasses three distinct sub-projects, each addressing specific aspects of AI education:

1. Games4AI: This sub-project focuses on theoretical exploration, incorporating concepts from Game Theory and Mechanism Design alongside gamification techniques. The primary objective is to develop an innovative pedagogical methodology for teaching AI principles to Generation Z students and individuals interested in delving into the realm of AI. Games4AI aims to leverage gamified approaches to make the learning process engaging and interactive.
2. SMaILE App: The SMaILE App constitutes a mobile application designed to introduce young learners to artificial intelligence through an edu-gaming experience. The app guides players on a playful, educational journey, where they assume the role of protagonists in constructing a sustainable SMAILE City. Powered by AI tools and engines, the SMaILE App provides an immersive and interactive platform to foster AI understanding among its users.
3. EmpAI: This sub-project aims to identify innate abilities in children and adolescents that, if nurtured, can enhance their comprehension of machines and systems exhibiting rational behaviors. Through EmpAI, the SMaILE project develops an educational platform and a range of ludic activities to stimulate computational skills and foster the acquisition of fundamental AI competencies

Fig. 3.3 View of the city in SMaILE app (2D and 3D)

in young learners. The focus is empowering students to develop computational thinking and engage with AI concepts hands-on and experientially.

The integrated structure of the SMAILE project, comprising the Games4AI, SMaILE App, and EmpAI sub-projects, enables a holistic exploration of AI education. The project aims to foster a comprehensive understanding of AI principles while nurturing the next generation of AI-literate individuals by incorporating theoretical foundations, interactive applications, and skill-building activities. Through collaborations with various stakeholders and leveraging cutting-edge methodologies, SMAILE seeks to revolutionize AI education and position Turin as a prominent hub for AI instruction and innovation.

Although the entire project has many different aspects, we will focus on the learning game app, the SMaILE app, and the related teacher training.

The SMaILE app [9], designed for lower and upper secondary school students, is a simulation game that aims to engage and educate young learners about AI principles using a constructionist pedagogical approach. The app utilizes game-based learning and gamification techniques to transform enjoyable activities into profound knowledge, fostering a deeper understanding of AI concepts. Through the city-building game (Fig. 3.3), students progressively construct a sustainable and livable smart city, applying AI concepts and principles in their decision-making processes. The game mechanics follow a constructionist approach, enabling users to experience, use, and link each tile within the city to specific learning outcomes, all contributing to the overarching mission of teaching AI principles to students.

The app incorporates various key features and functionalities that enhance students' learning experiences in AI. Firstly, the game mechanics are based on fundamental AI concepts, ensuring that each activity and service in the city corresponds to a specific mini-game that serves as a training ground for specific concepts:

- City building—constraint satisfaction problem
- Hospital—heuristic search
- Park selection—min-max algorithm
- Park cleaning—winning strategy
- Gardening—Game Theory
- Supermarket—-neural network and classification

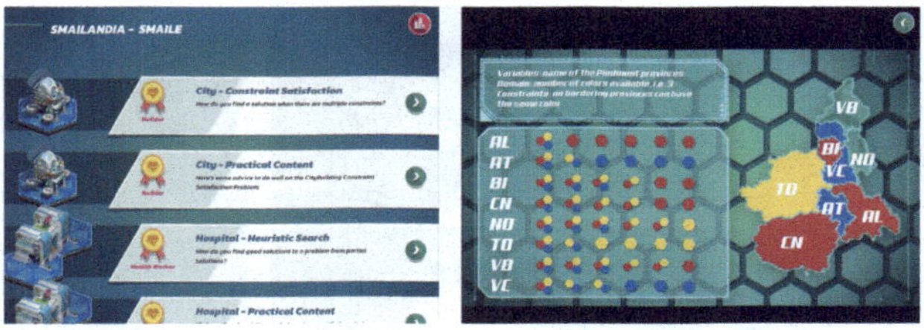

Fig. 3.4 Educational video in SMaILE app with an example

- Station—alpha zero and Monte Carlo technique
- School—reinforcement learning
- Cinema—natural language processes

Additionally, the app offers educational videos (see Fig. 3.4) for each activity or service in the city that provide formal explanations and guidance to students, further consolidating their understanding of AI principles.

Moreover, the SMaILE app incorporates gamification elements to maximize user engagement and motivation. Points, rewards, and virtual goods are employed to incentivize students to actively participate in the learning process and progress through the different levels and challenges. By gamifying the experience, the app creates an immersive and enjoyable environment where students can develop problem-solving skills, enhance their computational thinking abilities, and foster creativity. These elements provide a more engaging and interactive learning experience than traditional lecture-based approaches.

Incorporating the SMaILE app into the educational process yields several benefits, particularly in promoting student engagement and fostering a deeper understanding of STEM concepts, specifically in AI. The interactive and hands-on nature of the app encourages active participation, allowing students to explore various solutions and strategies to build and manage their smart cities. By linking AI principles to real-world applications within the game, students can witness the direct impact of their decisions and develop a holistic understanding of the subject matter. This not only enhances their knowledge but also cultivates soft skills such as critical thinking, problem-solving, and decision-making skills, all of which are crucial in the field of engineering and other STEM disciplines.

Our findings presented at the IEEE GEM conference [12], validated through GUESS-13 surveys [28] and focus groups, indicate that the app has been successful in engaging players aged 10–18, particularly females. The app's integration of AI concepts not only adds a layer of complexity but also serves pedagogical purposes, making it a unique addition to the educational gaming landscape. However, there are areas for improvement, particularly in graphics quality. Future work will aim to address these issues and expand the app's educational capabilities. The positive reception from committed players and statistical significance in gender preferences provide a promising avenue for the app's future development and potential impact in educational settings.

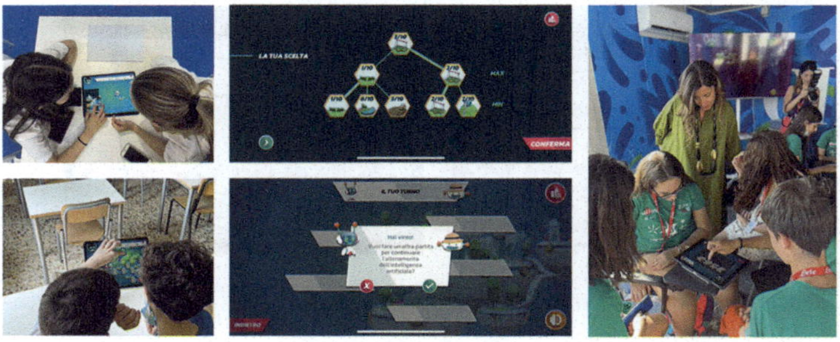

Fig. 3.5 Some pictures and screen example of the SMaILE app in action

Furthermore, the SMaILE app empowers students by giving them a sense of agency and ownership over their learning (Fig. 3.5). Through the gamified experience and the opportunity to create and customize their smart cities, students become active participants rather than passive recipients of knowledge. This sense of agency not only increases motivation but also fosters a more profound understanding of connection and investment in the learning process.

The teachers' training course run with the support of Ufficio Scolastico Regionale Piemonte, presents twofold objectives: to foster a deeper understanding of AI concepts and to empower young students to not only consume AI technology but also create it. Through a series of engaging sessions, the SMaILE teachers' training unfolds as a unique educational journey, integrating digital resources, teaching practices, unplugged activities, and student digital competence development. The SMaILE teacher training program operates within digital competence education (DigCompEdu), encompassing multiple focus areas:

- *Digital resources (area 2)*: The program taps into digital resources to explore innovative teaching methods and AI concepts.
- *Teaching and learning practices (area 3)*: Through hands-on activities, the training promotes interactive teaching and learning practices centered around AI education.
- *Student digital competence development (area 6)*: The training contributes to fostering students' digital competence by equipping educators with tools to facilitate meaningful AI learning experiences.

The program is tailored to novices with an entry level of A1, ensuring accessibility to educators at various stages of digital competence. The training spans 8 hours in total, providing an immersive and engaging educational experience.

The SMaILE teachers' training program is organized into four dynamic sessions, each delving into distinct aspects of AI education through engaging and practical activities:

Classification of Images and IoT Applications In this two-hour session, educators explore the classification of images as a means to convey narratives and construct applications involving the Internet of Things (IoT). The emphasis is on hands-on engagement, fostering an understanding of AI concepts through tangible experiences and storytelling. This activity bridges the digital and physical worlds, allowing educators to guide students in the classification of images while illuminating the concept of IoT. By manipulating real-world objects and images, students grasp the significance of AI's role in categorization and automation.

Autonomous Machines Through Unplugged Activities The second session, also spanning two hours, immerses educators in the realm of autonomous machines. Unplugged games like challenge 15 are employed to illustrate the concept of machines operating independently, shedding light on AI's role in creating self-sufficient systems. Teachers are guided through hands-on tasks that mirror the decision-making processes of machines, fostering an intuitive understanding of AI's capacity for independent operation.

AI Functionality: Research Systems and Winning Strategies The focus shifts to the functionality of AI. Educators delve into research systems and strategies that contribute to successful AI outcomes, enriching their understanding of AI's intricacies.

SMaILE App in the Classroom: Heuristics and Reinforcement Learning The final session revolves around the integration of the SMaILE app within classroom settings. Educators explore heuristic techniques and reinforcement learning, applying them to educational contexts through the app.

The SMaILE teachers' training program culminates in educators being equipped with the knowledge, tools, and strategies needed to foster meaningful AI education. By delving into the intricacies of AI through experiential learning, educators become transformative facilitators, inspiring young minds to engage with AI not just as consumers but as creators of technology. The program's structure reflects a commitment to cultivating digital competence, fostering innovative teaching practices, and shaping the next generation of digitally literate learners, poised to navigate the complexities of AI with enthusiasm and competence.

The SMaILE app serves as a dynamic gateway to AI education. Through interactive sessions, educators acquaint students with the app's functionalities, enabling them to explore AI concepts in an engaging and accessible manner. The app's gamified exercises create an immersive learning environment, ensuring concepts are not just understood but experienced. Leveraging the SMaILE app's features, educators craft gamified exercises that reinforce AI concepts. Through interactive challenges and scenarios, students navigate AI principles, solidifying their understanding while enjoying the educational journey. The SMaILE app opens avenues for dynamic discussions and explorations. Educators prompt students to critically analyze AI's role in various domains, fostering conversations that nurture curiosity and informed perspectives.

The integration of unplugged and plugged activities within the SMaILE teachers' training program underscores its comprehensive approach to AI education. Unplugged activities bring abstract concepts to life, grounding them in tangible experiences that cater to diverse learning styles. These activities lay the groundwork for plugged activities, where the SMaILE app emerges as a versatile tool for concept introduction, reinforcement, and discussion.

Through this integrated approach, educators cultivate AI enthusiasts who not only comprehend AI's intricacies but also possess the confidence to navigate its multifaceted landscape. The SMaILE teachers' training program empowers educators to guide students through meaningful AI learning experiences, leveraging both traditional and digital pedagogies to inspire a generation of technology-savvy learners. As educators harness the potential of unplugged and plugged activities, they spark curiosity, facilitate exploration, and set students on a trajectory of AI understanding that extends beyond the classroom.

3.1.2 Faculty Development for Higher Educational Lecturers

In the dynamic landscape of higher education, the role of university lecturers has evolved beyond conventional knowledge dissemination to encompass the cultivation of critical thinking, problem-solving, and collaboration skills among students. To effectively navigate these changing pedagogical demands, educators must continuously refine their teaching methodologies, embrace innovative approaches, and remain responsive to emerging educational paradigms. The "Provide Opportunity in Teaching" (POT) initiative developed at PoliTo represents a strategic effort to nurture the professional growth of higher education lecturers through diverse training opportunities and experiential engagements [3].

POT is a multifaceted framework designed to empower lecturers with a spectrum of pedagogical tools, experiences, and insights that extend beyond traditional classroom practices. As the educational landscape evolves, lecturers must adapt to changing learning needs and contribute to the transformation of higher education. In this context, POT acts as a catalyst, fostering an environment where educators can engage in continuous professional development, refine teaching skills, and exchange best practices.

POT emphasizes a holistic approach to faculty development, enabling educators to engage with pedagogical theories, innovative technologies, and novel methodologies. Through a variety of training activities, educators explore new frontiers of teaching, critically reflect on their instructional practices, and collaborate with peers to elevate the quality of higher education. From methodological training to leveraging advanced teaching technologies and embracing experiential learning, POT equips educators to navigate the present educational landscape and contribute to its evolution.

This comprehensive exploration of the POT framework elucidates its components, methodologies, and transformative potential. By championing continuous professional development, POT addresses the complex demands placed on modern higher education lecturers, allowing them to adapt and contribute to the evolving pedagogical ecosystem.

At the core of the POT framework lies methodological training, which immerses educators in diverse teaching methodologies, instructional strategies, and learner-centric approaches. This training enables educators to harmonize their instructional goals with the varied learning needs of their students, from flipped classrooms to project-based learning, fueling pedagogical innovation and aligning practices with the evolving educational landscape.

Two key offerings under this umbrella are the "English as a Medium of Instruction (EMI)" and the "Learning to Teach in Higher Education" courses. The EMI course fortifies the mastery of technical English for the transmission of scientific knowledge, fostering linguistic competency for seamless interaction with students and effective concept elucidation. The course includes one-on-one consultancy services to ensure personalized growth. The "Learning to Teach in Higher Education" course covers diverse themes such as student-centric perspectives, learning outcomes, didactic methodologies, active learning paradigms, and comprehensive learning assessment, providing educators with valuable teaching tools and insights [30].

In an era dominated by digital advancements, the assimilation of cutting-edge teaching technologies is imperative. The POT framework introduces educators to educational technologies, empowering them to integrate online platforms, virtual laboratories, simulation tools, and collaborative platforms into their teaching. This integration bridges the gap between traditional pedagogy and digital engagement, fostering personalized and impactful learning interactions.

The "Esperienze PoliTo" conference serves as a platform to disseminate best practices and monitor ongoing pursuits within the PoliTo community, highlighting strides in e-learning and MOOCs. The "Pillole Online" initiative enhances communication between the university and its student community, enabling educators to create impactful dissemination videos on scientific expositions and university services. The anticipated open education practice (OEP) courses add another layer of potency to these initiatives [24].

Experiential learning is the third dimension of the POT framework, emphasizing hands-on workshops, immersive seminars, and collaborative projects. These experiences encourage educators to experiment with novel teaching approaches, participate in peer mentoring, and share best practices. Methodological experiences stimulate reflective thought processes, enabling educators to fine-tune their pedagogical techniques based on real-world classroom dynamics and constructive feedback.

Dissemination conferences like "Costruire il futuro–da un'idea di Piero Angela," "TeatroScienza," and "GiovedìScienza" foster knowledge exchange and innovative dialogues, enhancing the educational landscape. Hybrid moments dedicated to the top 4% of students and mentoring of student teams further enhance knowledge exchange with scientific profundity.

The POT framework begins with a meticulous analysis of learners' distinct requirements, institutional objectives, and evolving demands within the engineering domain. This analysis informs the selection of appropriate teaching methodologies and strategies that align with desired learning outcomes. Regular assessments and feedback mechanisms empower educators to refine their approaches based on student performance and engagement.

Collaborative learning is central to the POT framework, cultivating a vibrant community of educators who participate in insightful discussions, workshops, and interdisciplinary dialogues. This network enriches educators' pedagogical repertoire, fostering the exchange of collective wisdom and enhancing teaching practices.

Introspective reflection is another key practice, where educators critically assess their teaching experiences and distill valuable lessons to inform future iterations. This reflective practice cultivates a growth mindset, compelling educators to adapt their strategies in response to evolving educational paradigms and learner needs.

The POT framework underscores the importance of continuous professional development in higher education. In a dynamic landscape, educators must impart technical knowledge and nurture competencies such as critical thinking, problem-solving, and adaptability. By engaging with the POT framework, educators amplify their teaching efficacy, aligning their practices with industry requisites and societal shifts.

Moreover, the POT framework represents a paradigmatic shift, positioning faculty development as a cornerstone of institutional excellence. This framework fosters a culture of lifelong learning within educational institutions, where faculty members embrace innovative pedagogy, engage in continuous skill enhancement, and contribute to the transformation of the education landscape.

In conclusion, the POT framework envelops educators in a realm of exploration, expansion, and excellence. The courses, initiatives, and experiences within this framework empower educators and orchestrate a symphony of educational evolution within higher education. Based on this extensive experience, a new training focusing on EE has been developed, the Teach2Teach program. It exemplifies comprehensive and impactful professional development, embracing learner-centered pedagogy, cross-cultural engagement, and community collaboration. As higher education evolves, lecturer training remains integral to nurturing educators who navigate modern education complexities and inspire future engineers and innovators.

3.1.3 Embracing Open Educational Practices (OEPs)

Open educational practices (OEPs) have emerged as a transformative force, reshaping the traditional paradigms of teaching, learning, and knowledge dissemination. The integration of digital technologies, coupled with the principles of openness and collaboration, has ignited a paradigm shift that extends far beyond the confines of conventional classroom settings. This subsection delves into the profound implications of OEP, exploring their significance within the realm of EE at PoliTo.

At its core, OEPs encompass a range of activities that leverage open principles and digital technologies to enhance the creation, sharing, and use of educational resources and knowledge. OEP encourages educators to collaborate, innovate, and engage in a dynamic exchange of ideas, transcending geographical and disciplinary boundaries. As educators harness the power of open licenses, online platforms, and collaborative methodologies, they unlock the potential to transform traditional education into a participatory and inclusive experience.

The relevance of OEP within higher education cannot be overstated. Today's higher education landscape is characterized by an unprecedented proliferation of information and a growing demand for flexible, accessible, and learner-centered educational experiences. Traditional pedagogical approaches, while invaluable, often face limitations in meeting these evolving demands. This is where OEP emerges as a catalyst for change.

By embracing OEP, institutions of higher learning can address several critical imperatives simultaneously:

Access and equity OEP breaks down barriers to education by providing open access to resources and knowledge. This democratization of education empowers learners from diverse backgrounds to engage with high-quality content, regardless of geographical location or socioeconomic status.

Innovation and collaboration OEP fosters a culture of innovation and collaboration among educators. It encourages the creation and sharing of open educational resources, enabling educators to co-design courses, experiment with novel teaching methodologies, and collectively advance pedagogical approaches.

Customization and engagement Through OEP, educators can tailor content to meet the unique needs of their students. This customization enhances student engagement and active learning, leading to deeper understanding and improved retention.

Professional development OEP offers educators opportunities for continuous professional development. By participating in open communities, educators can share experiences, learn from peers, and refine their teaching practices based on collective wisdom.

Global impact The open nature of OEP transcends institutional borders, enabling educators to contribute to the global knowledge commons. This interconnectedness facilitates the exchange of best practices and innovative ideas, driving educational advancement on a global scale.

OEPs have the potential to reshape education at PoliTo. By investigating the incorporation of OEP within the institution, we uncover not only the challenges and opportunities it presents but also the unique ways in which educators navigate the open landscape to enhance EE. The following exploration draws from an in-depth interview study conducted at PoliTo, shedding light on how OEP is being harnessed and its impact on EE within the institution.

The Open Educators Factory (OEF) framework, depicted in Fig. 3.6, serves as a comprehensive framework to assess educators' fluency in working with open approaches across various domains of their work. This framework encompasses four fundamental columns: design, content, pedagogy, and assessment [23]. Its aim is to offer a holistic view of the open educator's role, recognizing the diverse paths to openness and providing educators with a roadmap for their open journey.

Areas of activity

	A. Design	B. Content	C. Teaching	D. Assessment
Open Collaboration	**3. Open designer** Shares her course design ideas and curriculum openly through social media, including with colleagues and with students.	**3. Expert OER user** Re-shares resources she has reused openly through social media and OER repositories. Uses resources created by others. Searches for OER through social media and repositories. Shares links and resources beyond the classroom, through an open online identity.	**3. Open teacher** Encourages participation from non-enrolled students in her courses. Implements methods that foster co-creation of knowledge by students. Fosters students to contribute to public knowledge resources. Encourages learners to access freely available online content. Shares examples of teaching practice in open subject-related communities.	**3. Open evaluator** Uses open assessment practices such as peers assessment or e-portfolios. Engages communities of practices to assess students' work.
Second transition phase: becoming an Open Educator				
Bilateral collaboration/Small groups	**2. Collaborative designer** Collaborates in designing her courses with close colleagues, either from the same university or from international subject-related teams.	**2. Familiar with OER** Re-shares resources she has reused among close colleagues. Produces and share one's own resources under open licences. Reuses resources recommended by trusted people.	**2. Engaging teacher** Adopts seminars-like strategies, either offline or through restricted online spaces (Chats, Discussion forums). Uses "flipped-classroom" methodologies. Uses the university LMS, to share links and resources with the students of her courses.	**2. Innovative evaluator** Experiments with peers-based assessments methods.
First transition phase: becoming aware about openness in education				
Individual work	**1. Individual designer** Designs her courses on her own, based on her knowledge and experience.	**1.New to OER** Might use digital resources found on the web to enhance teaching and learning. Does not produce openly-licensed content.	**1. Traditional teacher** Adopts traditional trasmissive pedagogy	**1. Traditional evaluator** Uses traditional assessment methods such as tests or classwork.

Fig. 3.6 The open educators factory framework

The OEF framework is the culmination of an extensive literature review on open education. It draws from insights found in definitions, conceptual frameworks, and guidelines geared towards university educators aiming to enhance their open fluency. This framework was further refined through discussions with experts in the field of open education.

The OEF framework classifies educators into distinct profiles within each domain. In terms of design, three typologies emerge: the open designer, who shares course design ideas and curriculum openly through social media; the collaborative designer, who engages in collaborative course design with colleagues from the same or international subject-related teams; and the individual designer, who independently designs courses based on their knowledge and experience.

Within the resources domain, educators are categorized into profiles based on their engagement with open educational resources (OERs). The OER expert re-shares openly reused resources, uses resources created by others, and actively contributes to OER repositories. The OER novice creates and shares resources under open licenses, while the "New to OER" educator leverages digital resources from the web to enhance teaching.

In the realm of teaching, profiles are differentiated based on their approaches. The open teacher fosters co-creation of knowledge among students, encourages contribution to public knowledge resources, and shares teaching practices openly. The engaging teacher adopts interactive strategies, such as seminars and flipped-classroom methodologies, to enhance student engagement. The traditional teacher adheres to traditional transmissive pedagogy.

In the area of assessment, educators are grouped by their assessment methods. The open evaluator employs open assessment practices like peer assessment and e-portfolios, while the innovative evaluator experiments with peer-based assessment methods. The traditional evaluator relies on conventional assessment methods such as tests or classwork.

The OEF framework served as a foundational tool to guide an extensive study at PoliTo. A total of 173 teachers from various academic fields participated in the study by filling out a questionnaire on the OEF platform [24]. The participants comprised a diverse demographic, including different ranks, disciplines, ages, and gender distribution.

The findings illustrated a comprehensive snapshot of the openness capacity and gaps within the institution's teaching staff. Notably, the framework's holistic approach allowed for a nuanced understanding of openness across the dimensions of design, content, pedagogy, and assessment.

3.1.3.1 Learning Design

A substantial number of educators (116) collaborated with colleagues to design their courses. Seven educators were identified as open designers, who openly shared their ideas and curriculum through social media even before their courses commenced.

This early openness demonstrates a commitment to fostering an open attitude from the outset of the teaching cycle.

3.1.3.2 Teaching Resources

A significant majority (100) of respondents demonstrated awareness and use of OERs, including applying open licenses to their materials and sharing resources among colleagues. Notably, 22 educators were categorized as OER experts, showcasing their fluency in searching, adapting, and sharing resources openly through repositories and social media.

3.1.3.3 Teaching Methods

While traditional frontal teaching remained predominant (94 respondents), a notable proportion (80) engaged students through offline and online collaborative methods. Eight educators were classified as open teachers, who not only fostered co-creation of knowledge but also shared their teaching practices in open communities, indicating their readiness to lead by example.

3.1.3.4 Assessment

The majority of educators (162) adhered to traditional assessment methods, reflecting the established norm. However, a noteworthy group of 13 educators implemented innovative open assessment methods, signifying a willingness to experiment and embrace openness even in areas with normed practices.

Using this study [24] as starting point, an in-depth interview study was conducted [8]. This new study aimed to uncover the perspectives of educators regarding their understanding of OEP, its practical implementation, and the challenges and opportunities it presents within the university's unique setting.

The study revealed two distinctive groups among the interviewed educators: those who primarily perceive OEP as teaching with OERs and those who embrace a broader ethos of open teaching practices. The former group, referred to as Group A, emphasized the importance of organizational conditions, human resources, and collaboration with colleagues. These educators recognized how specific teaching obligations could serve as opportunities for innovation, particularly with technological tools such as the polling system provided by PoliTo.

On the other hand, educators in Group B saw OEP as a comprehensive approach that extends beyond OERs, where student engagement and collaboration play pivotal roles. Their teaching decisions were guided by interaction with students and the desire to enhance teaching practices progressively. This group identified challenges related to institutional incentives, resources, and the need for structurally avant-garde classrooms that support open teaching practices effectively.

Both groups highlighted the need for enhanced support from trained human resources in OEP and acknowledged the industry's role in fostering openness. Additionally, they expressed a shared interest in tailoring teaching approaches and resources to suit the needs of different student cohorts, leveraging the potential of OEPs to facilitate this process.

The findings of this study have substantial implications for the training opportunities and professional development of educators at PoliTo. The insights gained from educators in both Group A and Group B highlight the importance of fostering a culture of openness that transcends the use of resources alone. It is evident that OEPs can serve as catalysts for pedagogical innovation, enabling faculty members to collaboratively access each other's materials and approaches, thus enriching the teaching experience.

The analysis and insights derived from the study were pivotal in suggesting several actions to policymakers:

- Targeted professional development:

  Customized workshops Design workshops to address specific needs identified in the OEP study, ranging from introductory sessions on OEP fundamentals like copyrights to advanced training on creating and sharing OER.

  Peer mentoring programs Establish a peer mentoring program, pairing experienced educators proficient in OEP with those less familiar to facilitate knowledge transfer and collaborative learning.

- Resource development and sharing:

  Creation of OER repositories Develop centralized OER repositories to provide a platform for educators to share their materials, access resources created by colleagues, and collaborate on resource development.

  Incentives for OER contribution Introduce recognition awards, grants, and professional development credits to encourage faculty participation in creating and sharing OER.

  Technological infrastructure Make investments to upgrade the technological infrastructure to support OEP, including enhancing the learning management system (LMS) to integrate OER repositories and provide tools for collaborative resource creation.

- Enhanced collaboration and community building:

  OEP community of practice Establish an OEP community of practice (CoP) to facilitate ongoing collaboration among educators through regular meetings, seminars, and online forums.

  Cross-institutional partnerships Seek partnerships with other institutions to broaden the impact of OEP through collaborative projects and joint initiatives.

As we delve deeper into training opportunities and professional development, it is crucial to recognize the different perspectives and motivations within the educator community. By addressing the barriers identified by educators, such as the time required for OEP implementation and the need for tailored support, PoliTo can facilitate the integration of open practices into various educational contexts.

3.2 Course Development and Methodologies

3.2.1 The Role of Instructional System Design

Where the pursuit of knowledge intersects with the demands of rapidly evolving industries, the role of instructional design emerges as a guiding light to illuminate the path of effective learning. Instructional design, as a systematic and strategic approach, plays a pivotal role in shaping educational experiences that transcend traditional paradigms. This subsection delves into the heart of instructional design, elucidating its essence and exploring why it holds particular significance in the realm of EE.

At its core, instructional system design can be envisioned as the art and science of crafting intentional learning journeys. It involves the orchestration of diverse components, such as content, pedagogical strategies, assessments, and technologies, into a harmonious symphony that resonates with learners' cognitive processes. In this age of information abundance, instructional design acts as a discerning curator, meticulously selecting and arranging educational elements to foster engagement, comprehension, and mastery.

In EE, where the fusion of theory and application underpins the curriculum, instructional design emerges as a powerful ally. The intricacies of engineering concepts and practices demand an approach that not only imparts knowledge but also cultivates problem-solving prowess and innovation. This subsection navigates through the facets of instructional design methodologies, exploring how they bolster the effectiveness of teaching and learning within the engineering domain.

The pages ahead unravel the significance of instructional design in EE, drawing insights from scholarly contributions that illuminate its transformative potential. As Ng et al. [25] contend, instructional design principles possess the ability to enhance the pedagogical value of learning objects and activities, enabling learners to navigate the complexities of engineering disciplines with acumen and confidence. Little and Cardenas [19] further underscore the value of active learning methodologies, accentuating the symbiosis between hands-on design projects and the cultivation of design-oriented thinking.

Amidst the discourse, the essence of instructional design intertwines with the evolution of EE paradigms. Minichiello and Caldwell [20] explore the synergy between design-based research and the modern tools of technology, accentuating how instructional design methodologies can catalyze innovation and address contemporary challenges. Morris and McAdams [21], in turn, shine a light on the practical integration of design projects into engineering courses, fostering an immersive and experiential understanding of engineering design principles.

The application of instructional design methodologies has emerged as a pivotal approach to enhance the quality and effectiveness of teaching and learning processes. The significance of incorporating instructional design principles lies in their capacity to synergize pedagogical strategies with the distinct demands of EE. By systematically structuring course content and activities, these methodologies address the unique challenges posed by complex technical subjects and equip both educators and students with a comprehensive framework for success.

The role of instructional design methodologies extends beyond theoretical insights, as evidenced by Morris' practical insights into design integration methods. Such methods permit the seamless infusion of design projects into established engineering courses, empowering students to engage with the learning process actively. Notably, this active engagement aligns with Thom, Crossley, and Thom's work [34], which illustrates how structured engineering design tools can holistically transform curricula, ensuring their relevance and alignment with evolving industry needs.

In a broader context, the discourse on instructional design methodologies in EE has evolved to encompass various perspectives. Sukacke et al. [32] champion a paradigm shift from traditional teacher-centered education towards instructor-led instructional design, mirroring the trend towards active learning methods.

Collectively, these scholarly contributions underscore the pivotal role of instructional design methodologies in EE. Beyond mere theoretical significance, these approaches hold the potential to reshape the landscape of teaching and learning in engineering. Through the systematic alignment of pedagogical principles with the intricate demands of engineering disciplines, instructional design methodologies emerge as catalysts for enhanced educational experiences and improved learning outcomes.

In the realm of instructional system design, various systematic models have emerged to guide the creation of effective learning experiences. These models provide structured frameworks that streamline the process of developing instructional materials and strategies. Let us explore a selection of these widely recognized models, each offering unique approaches to crafting meaningful educational content.

3.2.1.1 ADDIE Model

The ADDIE model, a cornerstone of instructional design, offers a comprehensive approach to crafting impactful learning experiences. Standing for analysis, design, development, implementation, and evaluation, ADDIE provides a structured framework for systematically designing instruction. The process begins with a meticulous analysis of the learning context and target audience, facilitating the identification of learning objectives and needs. In the design phase, instructional strategies, content, and assessments are carefully crafted to align with these objectives. Development involves creating the actual instructional materials and resources, translating design concepts into tangible learning tools. The implementation phase brings these resources to learners, ensuring the instructional plan is put into action. Finally, the evaluation phase assesses the effectiveness of the instruction through formative and summative evaluations, allowing for iterative improvements and refinements. The ADDIE model's iterative nature fosters continuous enhancement, making it a versatile and widely employed framework in the field of instructional design.

3.2.1.2 Dick and Carey Model

This model emphasizes careful analysis, alignment, and evaluation of instructional elements. It provides a structured framework for designing effective learning experiences, ensuring that the instruction is well aligned with desired outcomes. It involves identifying instructional goals, conducting detailed task analysis, specifying performance objectives, designing assessments, selecting appropriate instructional strategies, and evaluating the instruction. Its structured approach helps ensure that instruction is well aligned with desired outcomes. The model comprises several stages that guide instructional designers through the process: instructional goals, learner and context analysis, task analysis, performance objectives, instructional strategy, instructional materials development, formative evaluation, and summative evaluation. The Dick and Carey model's structured approach ensures that the design process is thorough and aligns with the learning goals. By systematically analyzing the context, breaking down tasks, and aligning objectives, this model supports the creation of instruction that is effective, efficient, and meaningful for learners.

3.2.1.3 SAM Model (Successive Approximation Model)

The SAM model focuses on rapid prototyping and iterative design through collaboration. SAM deviates from the traditional linear design process and instead focuses on creating effective instruction through a series of iterations. The model is particularly well suited for courses where flexibility and responsiveness to learner needs are essential. It involves three main phases: preparation, iteration, and implementation. In the first phase, the design team collaborates with stakeholders, including subject matter experts and learners, to gather information about the learning goals, content, and desired outcomes. The team defines project constraints, identifies the target audience, and outlines the project's scope and objectives. The iteration phase involves a series of cycles, each comprising three steps: design, develop, and review. During the design step, the team creates a prototype of the instruction, focusing on key content and interactions. In the develop step, the prototype is developed into a functional instructional module. This module is then reviewed by stakeholders, including learners, to gather feedback. In the final phase, the instructional materials are refined based on the feedback received in the previous phase. The refined materials are then fully implemented for a wider audience. This phase can involve instructor training, learner orientation, and the actual delivery of instruction. The SAM model's iterative approach is its core strength. It allows for rapid testing and refining of instruction, ensuring that learner feedback and changing needs are promptly incorporated. This flexibility is particularly valuable when dealing with complex subjects, evolving content, or situations where the learning environment may change.

3.2.1.4 Gagne's Nine Events of Instruction

Gagne's Nine Events of Instruction is a widely recognized instructional design model developed by Robert Gagne in 1965. It outlines a sequence of nine events that enhance the effectiveness of learning experiences by engaging learners, promoting understanding, and facilitating retention. Each event serves a specific purpose in guiding learners through the learning process. It starts with gaining learners' attention, informing them of the learning objectives, presenting the content, providing guidance and practice, eliciting learner performance, providing feedback, assessing performance, and enhancing retention and transfer of knowledge.

3.2.1.5 ASSURE Model

This model provides a structured approach for planning and implementing effective instruction. Developed by Heinich, Molenda, Russell, and Smaldino, the model is designed to ensure that instructional materials and strategies are well aligned with learning objectives and learner needs. The acronym "ASSURE" stands for the steps involved in the process: analyze learners; state objectives; select methods, media, and materials; utilize media and materials; require learner participation; and evaluate and revise. The ASSURE model's strength lies in its systematic approach to instructional design. By emphasizing learner analysis, alignment with objectives, and active learner participation, the model ensures that instruction is learner-centered and effective. It also encourages the use of various media and materials to cater to diverse learning styles and preferences. The model's iterative nature, through the evaluation and revision step, promotes ongoing improvement in instructional design based on learner feedback and performance data.

3.2.1.6 ARCS Model

The ARCS model of motivational design, developed by John Keller, is an instructional design framework that focuses on enhancing learner motivation and engagement throughout the learning process. The model aims to create instruction that captures learners' attention, demonstrates the relevance of the content, builds their confidence, and ensures their satisfaction with the learning experience. The ARCS acronym represents the four key components of the model: attention, relevance, confidence, and satisfaction. The ARCS model provides a strategic framework for designing instruction that appeals to learners' intrinsic motivations and encourages active engagement. By addressing these four motivational factors, instructional designers can create learning experiences that are more enjoyable, meaningful, and effective. The model recognizes that motivated learners are more likely to stay engaged, retain information, and apply their learning beyond the instructional setting.

3.2.1.7 RID Model

The RID (rapid instructional design) Model is an instructional design approach that emphasizes efficiency and speed in the creation of instructional materials. It is particularly useful when there is a need for quick development of instruction due to time constraints, changing content, or other factors. The RID model streamlines the design process to expedite the creation of effective learning experiences. While not as detailed as some other instructional design models, RID focuses on key elements that contribute to rapid development: analysis, design, development, implementation, and feedback. The RID model is particularly suited for situations where immediate instruction is required, such as in rapidly changing work environments, emergency training, or short-term projects. It is not as comprehensive as other models but offers a pragmatic approach to developing instruction quickly while still maintaining a focus on learning outcomes. While speed is a key aspect, it is important to strike a balance between speed and ensuring that the instructional materials are effective and aligned with the desired outcomes.

These are some of the most popular models that offer structured frameworks for designing instruction and ensuring effective learning experiences. The choice of model depends on factors such as the nature of the content, the target audience, available resources, and the desired level of detail and planning.

Using the ADDIE model in EE offers several benefits that contribute to the effective design and delivery of instruction. The systematic approach of the ADDIE model aligns well with the complex and technical nature of engineering disciplines, resulting in improved learning outcomes and overall student satisfaction. As evidenced by Swain and Sahoo [33] and Peterson [27], the following benefits can be observed:

Enhanced learning effectiveness The ADDIE model's emphasis on thorough analysis, clear objectives, and aligned instructional strategies ensures that engineering content is presented in a structured and coherent manner. This approach promotes a deeper understanding of complex concepts and fosters critical thinking skills among engineering students.

Improved student satisfaction The systematic design process of the ADDIE model results in well-organized and engaging learning experiences. As a result, students are more likely to find the instruction meaningful, relevant, and enjoyable, leading to higher levels of satisfaction with the course.

Adaptation to technical content EE often involves intricate technical content that can be challenging to present. The ADDIE model allows instructors to break down complex topics into manageable modules, select appropriate instructional methods, and develop materials that cater to various learning styles.

Tailored instruction Through the analysis phase, the ADDIE model encourages instructors to understand the specific needs, prior knowledge, and learning preferences of engineering students. This enables them to customize the instruction to suit the diverse backgrounds and abilities of the learners.

Effective assessment strategies The ADDIE model's inclusion of assessment design ensures that engineering students' progress and understanding are regularly measured. This enables instructors to identify areas of improvement and provide targeted feedback to enhance learning.

Time management Engineering departments often have a rigorous curriculum, and the structured approach of ADDIE helps streamline the development and delivery of instructional materials, reducing the time needed to create effective courses.

Continuous improvement The ADDIE model's iterative nature supports continuous improvement. By evaluating the effectiveness of instruction through the evaluation phase, educators can gather feedback from students and make necessary adjustments for future iterations, resulting in ongoing enhancement of the course.

Alignment with industry standards The ADDIE model's structured approach aligns well with engineering industry practices, where systematic design, analysis, and evaluation are critical. Students who experience instruction designed using the ADDIE model are better prepared for the demands of the engineering field.

In summary, implementing the ADDIE model in EE offers a structured, adaptable, and effective approach to designing and delivering instruction. The model's systematic phases ensure that learning objectives are met, student satisfaction is high, and instructional materials are tailored to the technical and conceptual challenges of engineering disciplines.

3.2.2 Engaging Students in Course Redesign: ThinkLab

The "ThinkLab: IDEA (Instructional Design Elementary Application) Workshop" is a transdisciplinary teaching course that was designed and conducted within the University College of Merit "Collegio Universitario Renato Einaudi" in Torino, Italy. The workshop aimed to review and redesign the first-year engineering courses offered by PoliTo under the central theme of "resilience" during the academic year 2020/21 [6]. The workshop's instructional design is closely aligned with the ADDIE model. This model serves as a guiding framework to facilitate the course redesign process, incorporating the essential stages required for effective instructional design.

3.2.2.1 Analysis

In the context of the ThinkLab: IDEA Workshop, the analysis phase aligns with the initial theoretical moment of the workshop. During this phase, the instructors present and explain the ADDIE model to the participating students. Groups are formed, each linked to a specific first-year engineering course subject. This phase involves analyzing the needs of the course, understanding the learners' characteristics, evaluating the learning context, and considering the existing content and infrastructure. The Hodges et al. [17] framework is utilized to guide the analysis of needs, learners, context, environmental scan, infrastructure, and content-task relationship. This phase facilitates a comprehensive understanding of the course's current state and informs subsequent design decisions.

3.2.2.2 Design and Development Phases

The design and development phases of the ADDIE model align with the macro-level and micro-level design workshops within the ThinkLab: IDEA Workshop. The macro-level design workshop, spanning four hours, engages students in a guided ADDIE cycle. Students are provided with reference materials, including slides, guidelines, and course descriptions. This workshop is structured into three assignments: analysis (1 hour); design, development, and implementation (2 hours); and course description (1 hour). During the analysis assignment, groups systematically address questions related to instructional needs, learning objectives, interactions, skills, knowledge, and assessment strategies.

The design, development, and implementation assignment delves into defining the strategy and technological solutions for the course redesign. Students identify learning objectives, activities, interactions, required skills, knowledge, and assessments for each course week. The micro-level design phase focuses on the detailed preparation of specific topics to ensure alignment with the course description. Bates's [10] features, including outcomes, overview, reading, discussion, practice, assessment, and more, serve as a framework to structure the implementation.

3.2.2.3 Implementation and Evaluation Phases

The implementation and evaluation phases are realized through the workshop's various components, including the macro-level and micro-level design workshops and the final presentation. Students, guided by the ADDIE model, actively engage in role-play methodologies, experiencing the teaching and design process from the perspective of a lecturer. This experiential approach helps them understand the implications of their design decisions and fosters a more meaningful learning experience.

The ADDIE model's iterative nature is reflected in the workshop's progression, promoting continuous improvement. The evaluation of the ThinkLab: IDEA Workshop is facilitated through feedback and surveys provided by the participants. The workshop's impact on student learning experiences and the integration of resilience concepts into course redesigns can be assessed, providing valuable insights for further refinement.

The ThinkLab: IDEA Workshop's alignment with the ADDIE model demonstrates a systematic and comprehensive approach to instructional design. The workshop leverages the ADDIE framework's strengths by integrating its phases into various activities, role-playing, and collaborative design efforts. Through this alignment, the ThinkLab: IDEA Workshop effectively guides students through the process of designing resilient and effective course experiences for first-year engineering students, showcasing the adaptability and applicability of the ADDIE model in addressing contemporary educational challenges.

The innovative approach employed by the "ThinkLab: IDEA Workshop" lies in its active engagement of students as co-designers in the process of course redesign. Traditionally, instructional design has primarily been driven by educators and administrators. However, this workshop introduces a paradigm shift by placing the students at the forefront of the redesign process. By immersing students in the role of instructors and encouraging them to re-envision the curriculum from an educator's perspective, the workshop harnesses their fresh insights, diverse perspectives, and intimate understanding of their own learning needs. This student-centric approach resonates with contemporary pedagogical principles, empowering learners to take an active role in shaping their educational experiences. Moreover, the workshop cultivates essential skills such as critical thinking, collaboration, and problem-solving, which are indispensable in today's rapidly evolving educational landscape.

The results of the "ThinkLab: IDEA Workshop" have been successfully integrated into the first-year engineering courses at PoliTo, exemplifying the workshop's tangible impact on curriculum enhancement. The insights and innovative ideas generated by students during the workshop have informed a comprehensive redesign of the first-year courses, ensuring alignment with the central theme of resilience and best practices in instructional design.

3.2.3 *Enhancing Efficiency Through Competition: TEACH-GYM*

The significance of instructional system design in shaping the quality and outcomes of courses within the educational landscape cannot be overstated. The choice and application of appropriate teaching methodologies are instrumental not only in engaging and enlightening students but also in optimizing the efficiency and effectiveness of the overall educational process. In the relentless pursuit of enriching and impactful learning experiences, the integration of innovative teaching methods has become an imperative for educational institutions.

In the pursuit of elevating instructional practices and nurturing a culture of perpetual advancement, an innovative initiative has emerged—TEACH-GYM: Grow Your Methodology. This visionary framework heralds a transformative shift in how educational institutions approach the design of instructional systems. By amalgamating the principles of gamification with the aspiration for pedagogical excellence, TEACH-GYM seeks to not only elevate the efficiency and effectiveness of courses but also to foster a dynamic and captivating learning environment. The initiative forms a profound and enduring connection between PoliTo, a distinguished European university, and TTPU, a prominent institution in Central Asia. The distinctive educational needs and disparities between these two regions have catalyzed a stimulating contest of ideas, designed to revamp teaching methodologies and recalibrate course content within bachelor programs. On one front, the contest prompts the creation of more specialized courses that cater to the specific demands of undergraduate education in Uzbekistan. On the other, it paves the way for the transfer of expertise and methodologies to the Italian university.

At its core, TEACH-GYM symbolizes a collaborative endeavor aimed at reshaping the conventional teaching methodologies prevalent within educational establishments. It transcends conventional approaches by leveraging the potential of competition to stimulate innovation, endorse experimentation, and engender tangible enhancements in course design and delivery. This framework provides educators with a platform to innovate within their instructional approaches, experiment with novel teaching strategies, and cultivate a learner-centric environment in sync with the ever-evolving requirements of students and the educational domain.

TEACH-GYM is guided by several key principles and objectives:

Innovation through collaboration TEACH-GYM fosters a sense of collaboration among educators and instructional designers. By encouraging the exchange of ideas and experiences, the competition creates an environment where innovative teaching methodologies can flourish.

Student-centered approach The competition places students at the forefront. Participants are challenged to develop teaching strategies that cater to diverse learning styles, promote active participation, and enhance overall comprehension and skill acquisition.

Continuous improvement TEACH-GYM underscores the importance of continuous improvement in education. Participants are inspired to critically assess existing teaching methods and propose enhancements that adapt to the evolving needs of learners.

Impactful learning outcomes The ultimate goal of TEACH-GYM is to enhance learning outcomes. Educators are motivated to design methodologies that not only impart knowledge effectively but also foster critical thinking, problem-solving skills, and a passion for lifelong learning.

The inaugural TEACH-GYM call witnessed remarkable participation and outcomes. Italian lecturers teaching at TTPU actively engaged in the initiative, submitting proposals to reshape their courses to meet the educational demands of the region. The results were promising: a total of 23 proposals were received, each

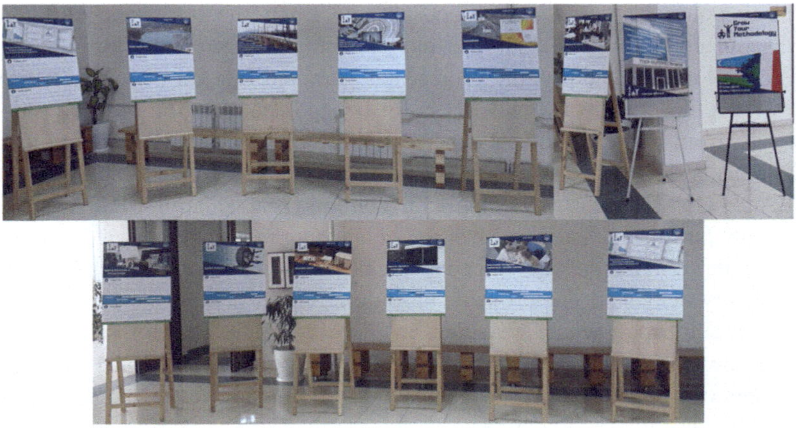

Fig. 3.7 Poster section of the winning proposals during the celebration for the 10th year of TTPU

reflecting innovative adaptations of instructional methods tailored to the diverse learning preferences of students. These proposals underscored the commitment of educators to enhancing the educational experience and achieving impactful learning outcomes (Fig. 3.7).

Furthermore, TEACH-GYM introduced a call inviting Italian lecturers at TTPU to review and reshape their courses in the context of the distinct educational environment. This initiative exemplifies the commitment to cross-cultural collaboration and adaptive teaching practices, ensuring that the courses remain responsive to the evolving academic landscape and the diverse needs of learners. Informed by the shared realization that at TTPU merely 10% of students choose to pursue Master of Science degrees and that students often juggle part-time jobs alongside their studies due to the cost of graduate education, TEACH-GYM propels a rethinking of pedagogical methods. Recognizing the necessity of adapting teaching methodologies to meet these multifaceted demands, TEACH-GYM presents an arena of change where innovation flourishes.

The inception of TEACH-GYM [1] saw initial trials in courses employing experiential learning and simulations, yielding remarkable outcomes. The adoption of these innovative methods led to a notable increase in student attendance and engagement, alongside elevated success rates and academic achievements. These encouraging results provided the foundation for the project's expansion, driving forward the vision of reshaping courses and instilling job-oriented skills.

The impact of TEACH-GYM extends beyond student outcomes, permeating the academic fabric of both institutions. The initiative bridges PoliTo and Turin Polytechnic University in Tashkent (TTPU) in a symbiotic manner. While PoliTo gains insights for refining its courses based on concrete feedback from TTPU implementations, the Uzbekistan institution benefits from enhanced teaching methodologies and invaluable experiential learning for its faculty members.

The response to the call for proposals underlines the resonance of TEACH-GYM's objectives, with 14 submissions spanning diverse scientific domains. A testament to its adaptability, the project has garnered interest from various disciplines, reflecting a collective eagerness for pedagogical reinvention. The range of proposals, from interdisciplinary experiential projects to focused teaching method enhancements, emphasizes a comprehensive approach to course transformation.

TEACH-GYM's trajectory hinges on continuous evaluation, and preliminary insights are promising. The revitalization of teaching methodologies at both institutions signifies a dynamic shift in pedagogical paradigms, fostering skills that align with industry demands. The project's unfolding impact heralds the emergence of a win-win relationship, where academic institutions collaborate for mutual growth and educational excellence. As TEACH-GYM continues to unfold, the ongoing implementation of innovative strategies promises a landscape where education transcends conventional boundaries.

Among the winning proposals that emerged, one that stands out is the creation of the REAL Lab, an acronym for Remote Automatic Control Laboratory [5], which embodies a remarkable collaborative endeavor with far-reaching implications for EE. It represents an innovative and collaborative initiative that bridges geographical distances to provide students with valuable educational opportunities. The lab's inception and its subsequent impact from the perspective of the lecturer highlight the dedication and transformative teaching approach that went into its creation. Its purpose was to establish a remote platform accessible to students, particularly those from Uzbekistan, enabling them to engage in practical learning experiences in the field of automatic control (further details on the students' point of view are available in the Sect. 2.9).

From the lecturer's point of view, the REAL Lab signified a significant shift in educational paradigms. It required the exploration of novel pedagogical approaches to effectively deliver content and facilitate experiential learning in a virtual environment. The lecturer's role evolved from being a traditional source of knowledge to a facilitator of interactive and hands-on learning experiences. This transformation demanded not only technical expertise but also the ability to adapt to the dynamics of remote learning.

The REAL Lab's establishment called for a multidisciplinary approach, necessitating collaboration between educators, engineers, and technology specialists. Overcoming technological barriers, designing remote experiments, and ensuring seamless communication across continents were among the challenges that the lecturer, alongside their collaborators, faced. The lecturer's commitment to this endeavor was driven by the belief in the educational value it could offer to students, as well as the potential to enhance cross-cultural exchange and collaboration.

The development and implementation of the REAL platform required a phased approach, involving iterative design, testing, and refinement. The collaborative effort between TTPU and PoliTo ensured that the platform's design and functionalities aligned with the educational objectives and pedagogical practices of both institutions. This joint effort fostered a strong sense of ownership and investment from the educators and students, creating a shared commitment to the success and continuous improvement of the remote laboratory.

By collaborating with TTPU and extending the REAL platform to Italian courses, PoliTo demonstrated its commitment to promoting internationalization in EE. The collaboration not only enriched the educational experience for students at both institutions but also fostered a deeper understanding and appreciation of different cultural and educational contexts. Furthermore, the partnership between TTPU and PoliTo served as a model for future collaborations, emphasizing the value of knowledge exchange and cooperation in enhancing EE globally.

The REAL Lab's impact was felt in multiple dimensions. It not only empowered Uzbek students with access to resources and learning opportunities that were previously unavailable but also enriched the educational experience of students at PoliTo. The lab's success exemplified the potential of technology-mediated education to transcend geographical limitations and create a global learning community.

This collaborative effort marked a significant step forward in leveraging technology for education, fostering intercultural understanding, and nurturing a sense of global citizenship among students. As the REAL Lab continues to evolve, its legacy serves as an inspiration for educators and institutions worldwide to explore innovative ways to enhance education and create meaningful connections across borders.

3.2.4 Innovation in Course Delivery: Some Examples

The integration of innovative teaching approaches, including the incorporation of remote labs, has marked a transformative shift in lecturers and students' life. As traditional educational paradigms continue to evolve, educators are exploring dynamic methods to enhance student learning experiences and foster deeper engagement with complex concepts. This evolution is particularly evident through initiatives like the REAL Lab, where the convergence of advanced technology and pedagogy is revolutionizing the educational landscape. By seamlessly combining theoretical knowledge with practical applications, remote labs offer students a unique opportunity to engage in hands-on experiments and collaborative learning, transcending geographical boundaries and traditional classroom limitations.

The REAL project stands as a groundbreaking testament to the transformative potential of remote labs in EE. By seamlessly amalgamating theory with practice, transcending geographical constraints, and fostering collaborative global learning, this initiative illuminates the path forward for modern pedagogy. As we delve further into the exploration of innovative teaching approaches, the REAL project serves as an emblem of how technology and education can converge to revolutionize the very essence of learning.

Another offspring of the TEACH-GYM initiative, which has yielded transformative results, is the revitalized Structural Analysis course at PoliTo [2, 4]. Inspired by the insights garnered from the TEACH-GYM experience, the lecturer spearheading the course undertook a comprehensive redesign, driven by a resolute commitment to aligning the curriculum with the dynamic demands of the contemporary educational landscape. The endeavor to infuse the course with the spirit of student-centered pedagogy, sparked by TEACH-GYM's ethos, led to a comprehensive redesign that bore testament to the lecturer's unwavering commitment to fostering enhanced learning outcomes. This section delves into the profound redesign of the Structural Analysis course, underscored by the integration of problem-based learning (PBL) principles, and navigates through the challenges encountered in this pedagogical transformation.

The lecturer's participation in the TEACH-GYM initiative served as a pivotal catalyst that kindled a fervent desire to extend the boundaries of innovation within the educational landscape. Encountering the synergistic blend of interactive teaching methodologies, the lecturer gained a profound appreciation for the transformative potential of student-centered pedagogy. Inspired by the rejuvenating spirit of TEACH-GYM, the lecturer envisioned a similar pedagogical metamorphosis for the traditional Structural Analysis course, aimed at fostering deeper engagement and more comprehensive understanding among students.

Central to the overhaul of the course was the integration of PBL, a pedagogical approach known for its ability to cultivate critical thinking, collaboration, and real-world problem-solving skills. Recognizing the need to bridge the gap between theoretical concepts and their practical applications, the lecturer skillfully infused authentic, real-world challenges into the curriculum. Through PBL, students were not merely passive recipients of knowledge but active participants in constructing their learning journey (see further details in the students perspective in Sect. 2.10).

Undoubtedly, the transition from a traditional instructional model to a PBL-infused curriculum was not without its challenges. One prominent hurdle was the need to strike a delicate balance between structured content delivery and the open-ended, exploratory nature of PBL. The lecturer grappled with aligning the learning objectives with the dynamic nature of project-based exploration, ensuring that while students delved into real-world problems, they remained firmly grounded in theoretical foundations.

Moreover, the heterogeneity of the student cohort posed both opportunities and challenges. Tailoring PBL experiences to cater to diverse learning styles and prior knowledge levels necessitated a flexible approach to instruction. The lecturer adeptly employed various scaffolding strategies, leveraging peer collaboration and individualized guidance, to ensure that every student could derive meaningful insights from the PBL process.

In conclusion, the reconstruction of the Structural Analysis course exemplifies the lecturer's bold initiative to revolutionize education. Guided by the spirit of TEACH-GYM and fortified by the principles of PBL, the course's redesign stands as an opportunity to the educator's resolute commitment to nurturing the holistic development of students. Despite challenges, the lecturer's dedication to navigating uncharted pedagogical terrain paved the way for a more enriching, relevant, and transformative learning experience within the context of structural analysis education.

In summation, the role of university lecturers in shaping the future of EE is both pivotal and transformative. As highlighted throughout this chapter, the integration of innovative pedagogical strategies, such as problem-based learning and the incorporation of advanced technological tools, significantly enhances the educational experience. Moreover, the continuous professional development of educators, supported by frameworks like TEACH-GYM and POT, fosters a dynamic and adaptive teaching environment. By embracing open educational practices and collaborative methodologies, lecturers can further democratize education, ensuring access and equity for all students. The practical implications are substantial: lecturers benefit from enhanced teaching resources, collaborative teaching practices, and professional recognition, while students gain access to customized learning materials and collaborative environments that enhance engagement and retention. Moving forward, the commitment to these innovative practices will not only enrich the learning journey of students but also prepare them to meet the evolving challenges of the engineering field with confidence and competence. The pursuit of excellence in EE, guided by these principles, will undoubtedly pave the way for a future where both educators and students thrive in a synergistic and ever-evolving academic landscape.

References

1. Ballatore MG, Stievano IS, Tabacco AM (2019) TEACH-GYM: grow your methodologies. In: Innovations, Technologies and Research in Education. Linda Daniela – University of Latvia, Latvia, pp 180–191. https://doi.org/10.22364/atee.2019.itre
2. Ballatore MG, Barpi F, Tabacco A (2020) Work-in-progress: pedestrian bridge application in a fundamentals of structural analysis course inside an architecture bachelor program. In: IEEE Global Engineering Education Conference, EDUCON. IEEE Computer Society, Washington, DC, pp 41–45. https://doi.org/10.1109/EDUCON45650.2020.9125140
3. Ballatore MG, Felisatti E, Montanaro L, Tabacco A (2020) TEACH-POT: provide opportunities in teaching. Excell Innovat Teach Learn 5(1):79–91. FrancoAngeli Edizioni. https://doi.org/10.3280/exioa1-2020oa10078. Retrieved from https://www.francoangeli.it/riviste/SchedaRivista.aspx?IDArticolo=66341
4. Ballatore MG, Barpi F, Crocker D, Tabacco A (2021) Pedestrian bridge application in a fundamentals of structural analysis course inside an architecture bachelor program. Int J Eng Pedagogy 11(5):4–18. International Association of Online Engineering (IAOE). https://doi.org/10.3991/ijep.v11i5.20151. Retrieved from https://doi.org/10.3991/ijep.v11i5.20151
5. Ballatore MG, Razza V, Regruto Tomalino D, Stievano IS, Tabacco AM (2022) The educational benefit of a remote automatic control laboratory. A win-win collaboration between Asia and Europe. Educ Knowl Soc 23. Universidad de Salamanca. https://doi.org/10.14201/eks.28495. Retrieved from https://revistas.usal.es/index.php/eks/article/view/28495
6. Ballatore MG, Stievano IS, Tabacco AM (2022) Think lab: we have an IDEA (instructional design elementary application). Irish J Acad Pract 10(1). Technological University Dublin. https://doi.org/10.21427/f8xg-ry19. Retrieved from https://arrow.tudublin.ie/ijap/vol10/iss1/8/

7. Ballatore MG, Romano B, Tabacco A (2023) Enhancing primary teachers' spatial ability to improve math learning: the MATABì project. In: Technological Ecosystems for Enhancing Multiculturality 2023 (TEEM23). (Accepted)
8. Ballatore MG, Nascimbeni F, Burgos D, Tabacco A (2023) Multiple roads towards openness: exploring the use of open educational practices within an Italian university. Italian J Educ Technol 31:89–100
9. Ballatore MG, Bernardini S, Como G, Damonte L, Franco Aixela S, Neascu A, Vanelli M (2023) SMaILE-App: a mobile game application to teach the principles of artificial intelligence to the z-generation. In: School Children and the Challenge of Managing AI Technologies: Fostering a Critical Relationship through Aesthetic Experiences, Routledge. Series: Routledge Research in Early Childhood Education. (In press)
10. Bates T (2016) The 10 fundamentals of teaching online for faculty and Instructors. Contact North, Canada
11. Dornian K, Moshirpour M, Behjat L (2020) K-12 digital skills programs as preparation for engineering study: a systematic literature review. In: 2020 ASEE Virtual Annual Conference Content Access Proceedings. ASEE Conferences. https://doi.org/10.18260/1-2--34889
12. Franco S, Ballatore MG, Bernardini S, Como G, Damonte L, Neacsu A, Vanelli M (2024) Unlocking AI literacy: the SMaILE-app gaming experience. In: 2024 IEEE Gaming, Entertainment, and Media Conference (GEM), pp 1–6. https://doi.org/10.1109/GEM61861.2024.10585394
13. Genalo L (2003) Engineering as context for K 12 science, mathematics, and technology education. In: 2003 Annual Conference Proceedings. ASEE Conferences. https://doi.org/10.18260/1-2-11449
14. Ghalia MB, Carlson R, Estrada V, Huq H, Ramos J (2016). Engaging K-12 teachers in engineering through a professional development program: implementation strategies, results and lessons learned. In: 2016 IEEE Frontiers in Education Conference (FIE). IEEE. https://doi.org/10.1109/FIE.2016.7757530
15. Guzey SS, Tank K, Wang H-H, Roehrig G, Moore T (2014). A high-quality professional development for teachers of grades 3-6 for implementing engineering into classrooms. School Sci Math 114(3):139–149. Wiley. https://doi.org/10.1111/ssm.12061
16. Hattie J (2008) Visible Learning: A Synthesis of Over 800 Meta-Analyses Relating to Achievement, 1st edn. Routledge, Milton Park. https://doi.org/10.4324/9780203887332
17. Hodges CB, Moore SL, Lockee BB, Bond MA, Jewett A (2020) An instructional design process for emergency remote teaching. In: Radical Solutions for Education in a Crisis Context. Lecture Notes in Educational Technology. Springer, Singapore, pp 37–52
18. Jemutai S, Webb P (2019) Effects of a 6 Brick Duplo Block guided play intervention on preliterate learners' visual perception. South Afric J Childhood Educ 9:1–8
19. Little P, Cardenas M (2001) Use of "Studio" methods in the introductory engineering design curriculum. J Eng Educ 90(3):309–318. Wiley. https://doi.org/10.1002/j.2168-9830.2001.tb00610.x
20. Minichiello A, Caldwell L (2021) A narrative review of design-based research in engineering education: opportunities and challenges. Stud Eng Educ 1(2):31. Virginia Tech Libraries. https://doi.org/10.21061/SEE.15
21. Morris E, McAdams D (2015) A review of practical design integration methods for existing engineering curriculum. In: 2015 ASEE Annual Conference and Exposition Proceedings. ASEE Conferences. https://doi.org/10.18260/p.23441
22. National Research Council (2009) Engineering in K-12 Education: Understanding the Status and Improving the Prospects. The National Academies Press, Washington, DC. https://doi.org/10.17226/12635. Retrieved from https://nap.nationalacademies.org/catalog/12635/engineering-in-k-12-education-understanding-the-status-and-improving
23. Nascimbeni F, Burgos D (2016). In search for the open educator: proposal of a definition and a framework to increase openness adoption among university educators. Int Rev Res Open Distance Learn 17(6). https://doi.org/10.19173/irrodl.v17i6.2736

24. Nascimbeni F, Burgos D, Campbell LM, Tabacco A (2018) Institutional mapping of open educational practices beyond use of Open Educational Resources. Distance Educ 39(4):511–527. Routledge. Retrieved from https://doi.org/10.1080/01587919.2018.1520040
25. Ng J, Lei C-U, Lau E, Lui K-S, Lam KH, Kwok TT, Hu X, Warning P, Tam V (2019) Applying instructional design in engineering education and industrial training: an integrative review. In: 2019 IEEE International Conference on Engineering, Technology and Education (TALE). IEEE. https://doi.org/10.1109/TALE48000.2019.9225920
26. Olds S (2004). Developing a standards based K 12 engineering curriculua through partnerships with university students and industry. In: 2004 Annual Conference Proceedings. ASEE Conferences. https://doi.org/10.18260/1-2--12987
27. Peterson C (2003). Bringing ADDIE to life: instructional design at its best. J Educ Multimed Hypermedia 12(3):227–241. Association for the Advancement of Computing in Education (AACE)
28. Phan MH, Keebler JR, Chaparro BS (2016). The development and validation of the game user experience satisfaction scale (guess). Human Factors 58(8):1217–1247
29. Porter T, West ME, Kajfez RL, Malone KL, Irving KE (2019) The effect of teacher professional development on implementing engineering in elementary schools. J Pre-College Eng Educ Res 9(2). Purdue University (bepress). https://doi.org/10.7771/2157-9288.1246
30. Serbati A, Felisatti E, Da Re L, Tabacco A (2018) Qualifying faculty teaching competences. The pilot experience at Politecnico di Torino. Form@re - Open J per la formazione in rete 18(1):39–52. https://doi.org/10.13128/formare-22548. Retrieved from https://oaj.fupress.net/index.php/formare/article/view/3715
31. Singer JE, Ross JM, Jackson-Lee Y (2016). Professional development for the integration of engineering in high school STEM classrooms. J Pre-College Eng Educ Res 6(1). Purdue University (bepress). https://doi.org/10.7771/2157-9288.1130
32. Sukacke V, Guerra AOPdeC, Ellinger D, Carlos V, Petroniene S, Gaiziuniene L, Blanch S, Marba-Tallada A, Brose A (2022) Towards active evidence-based learning in engineering education: a systematic literature review of PBL, PjBL, and CBL. Sustainability 14(21):13955. MDPI AG. https://doi.org/10.3390/su142113955
33. Swain A, Sahoo A (2020) Successful teaching learning method with satisfied and effective student - an ADDIE model approach. Int J Creat Res Thoughts (IJCRT) 8(6):1916–1922. Retrieved from http://www.ijcrt.org/papers/IJCRT2006260.pdf
34. Thom M, Crossley W, Thom JM (2002) The application of structured engineering design methodologies to engineering curriculum development. In: 32nd Annual Frontiers in Education, vol 3. https://doi.org/10.1109/FIE.2002.1158610
35. Wai J, Lubinski D, Benbow CP (2009) Spatial ability for STEM domains: Aligning over 50 years of cumulative psychological knowledge solidifies its importance. J Educ Psychol 101(4):817–835
36. Yoon SY (2011) Psychometric properties of the Revised Purdue Spatial Visualization Tests: Visualization of Rotations (The Revised PSVT:R). (Doctoral dissertation). Purdue University

Chapter 4
Discussion and Conclusion

> *The fishermen know that the sea is dangerous and the storm terrible, but they have never found these dangers sufficient reason for remaining ashore.*
>
> — *Vincent van Gogh, Letter to Theo van Gogh 14/05/1882*

Abstract This chapter discusses the transformative role of engineering education researchers (EERs) at Politecnico di Torino (PoliTo), highlighting their impact on both students and lecturers. Key contributions include pedagogical innovations, curriculum enhancements, and effective assessment methods. The chapter underscores the significance of spatial ability training, addressing prerequisite mismatches, and fostering a student-centered approach in engineering education. The integration of the scientific method and active learning strategies are emphasized to cultivate critical thinking and problem-solving skills. Additionally, the chapter explores the duality of teaching and research, advocating for the professional development of lecturers through international collaborations and innovative teaching methodologies. Practical recommendations for educators include adopting project-based learning, gamification, and technology integration to enhance the learning experience. The findings reveal the need for formal recognition of EERs within the Italian academic framework and the importance of continuous improvement in teaching practices. This chapter serves as a comprehensive guide for educators, administrators, and policymakers aiming to advance engineering education through evidence-based strategies and collaborative efforts.

The role of Engineering Education Researchers (EERs) extends far beyond the boundaries of traditional teaching and research. Their commitment is profound, aiming to elevate the educational experience for both students and faculty members. The central question guiding this pages has been: "What role does an EER play within an esteemed Italian technical university like Politecnico di Torino (PoliTo), and how does their work positively impact both students' understanding and lecturers' roles?" This research inquiry has served as a compass throughout our exploration of the multifaceted contributions of EERs.

M. G. Ballatore, *Empower Education to Foster Engineering*,
PoliTO Springer Series, https://doi.org/10.1007/978-3-031-75895-9_4

As the author of this book, I had the privilege of serving as the inaugural EER at PoliTo. Supported by esteemed full professors, I embarked on a journey to function as an academic scholar in the field of engineering education (EE). My role transcended disciplinary boundaries, bridging the gap between pedagogical innovation and engineering practice. Through rigorous research endeavors, I sought to scrutinize and refine pedagogical methods, curriculum design, and the overall learning experience within the realm of engineering.

The initiatives documented in these pages have had a profound impact on EE, both globally and specifically at PoliTo. These efforts contribute significantly to the advancement of EE by fostering innovative teaching methodologies, integrating technology, and promoting open educational practices. The emphasis on student-centered learning and the continuous professional development of lecturers have created a dynamic and adaptive educational environment.

From the implementation of various projects, several key lessons have emerged:

- **Pedagogical innovation**: Collaborating tirelessly, I continuously sought innovative teaching approaches and strategies to enhance the effectiveness of EE. Together with fellow educators, we explored novel methodologies, technologies, and active learning techniques that resonated with the ever-evolving needs of our students.
- **Curriculum enhancement**: Active participation in curriculum development and improvement was essential. We endeavored to ensure that our engineering programs aligned seamlessly with industry demands, emphasizing interdisciplinary learning and incorporating contemporary knowledge and skills.
- **Assessment and evaluation**: In the role of an EER, I learned to rigorously evaluate the outcomes of educational initiatives. Employing evidence-based assessment practices, I gauged student learning, retention, and success, facilitating data-driven improvements in our educational strategies.
- **Faculty development**: My engagement with lecturers and professors was marked by providing unwavering support and valuable resources to enhance their teaching capabilities. We carefully planned and facilitated professional development opportunities, fostering a culture of continuous improvement among our esteemed educators.

However, challenges such as aligning new teaching methods with traditional curricula and ensuring equitable access to resources remain.

Notably, the inception of the EER role at PoliTo marked a transformative period. Building a robust network of PoliTo professors and lecturers was pivotal, as it allowed us to bridge theory and practice effectively. Table 4.1 presents a comprehensive overview of the direct interactions and collaborations established by the EER. The data is organized by academic departments within PoliTo and includes the number of individuals reached directly by the EER, their respective career stages, and the count of publications with co-authors associated with each department. The EER's co-authorship is not included in the count for her department. This collaborative spirit within our academic community was instrumental in applying the theoretical role that an EER can play in practice. Simultaneously, the support

Table 4.1 Interactions and collaborations within the engineering education researcher's academic network (*) the total number of publications with at least one co-author affiliated with PoliTo. It is important to note that this count does not represent the sum of the row above, as a single publication may involve co-authors from multiple departments within the institution. Legend: RF, research fellow; L, lecturer; RTD, researcher; PA, associate professor; PO, full professor

Department	People directly reached						Number of publications with at least one co-author
	RF	L	RTD	PA	PO	Total	
DAD—Dip. Architettura e Design					2	***2***	1
DAUIN—Dip. Automatica e Informatica				1		***1***	1
DENERG—Dip. Energia				1	2	***3***	
DET—Dip. Elettronica e Telecomunicazioni				3	4	***7***	5
DIATI—Dip. Ingegneria dell'Ambiente, del Territorio e delle Infrastrutture			1	2	2	***5***	
DIGEP—Dip. Ingegneria Gestionale e della Produzione			1		1	***2***	1
DIMEAS—Dip. Ingegneria Meccanica e Aerospaziale				2	2	***4***	
DISAT—Dip. Scienza Applicata e Tecnologia				5	1	***6***	5
DISEG—Dip. Ingegneria Strutturale, Edile e Geotecnica				2		***2***	2
DISMA—Dip. di Scienze Matematiche "G. L. Lagrange"	2	2		6	5	***15***	20
Totale	***2***	***2***	***2***	***22***	***19***	***47***	***20****

and insights from the international EER community provided valuable guidance, shaping our understanding of this innovative role within the academic landscape.

Quantitative data collected from these initiatives indicate significant improvements in student performance metrics. For instance, students participating in the Matabì project showed enhanced spatial abilities and reduced math anxiety, leading to higher engagement in STEM subjects. Feedback from both students and educators highlights the positive impact of these projects on learning outcomes and teaching practices.

Qualitative reflections from lecturers and students provide deeper insights into the transformative effects of these initiatives. Testimonials from students reveal increased motivation and confidence in tackling engineering challenges, while lecturers report a greater sense of fulfillment and professional growth from adopting innovative teaching methods. These personal perspectives underscore the broader impact of these efforts on the educational community.

4.1 Fostering Engineering Excellence: Nurturing Student Potential

This section examines how EERs enhance student potential by focusing on spatial ability, addressing prerequisite mismatches, and promoting the scientific method. We emphasize the importance of a student-centered approach in education, which tailors teaching methods and curricula to meet students' needs, fostering holistic development and preparing them for success in engineering.

Central to the essence of education is the fundamental truth that students are the driving force behind the roles and responsibilities of lecturers. Without the presence of students, the very purpose and significance of lecturers cease to exist. It is in this recognition that we firmly advocate for the paramount importance of adopting a student-centered approach in education, particularly within the realm of technical subjects. Placing students at the heart of the educational process not only acknowledges their pivotal role but also underscores the essential nature of tailoring teaching methods and curricula to meet their needs, aspirations, and intellectual growth. In embracing the student-centered approach, we aim to foster an educational environment that prioritizes the holistic development of learners, empowering them to excel in their academic journey and contribute meaningfully to the field of engineering.

The essence of education can be traced back to its Latin origins in the word 'educere,' meaning to lead out or bring forth. This fundamental etymology holds a profound truth for the entire education system. It reminds us that education's overarching purpose is to guide and facilitate the unfolding of students' innate skills, talents, and aptitudes. As such, the education system should be a beacon of support, nurturing an environment where students are empowered to discover, harness, and express their unique potential. It is in this transformative process of "leading out" that education fulfills its most profound and noble mission—to enable individuals to illuminate their talents, contribute meaningfully to society, and embark on a journey of lifelong learning and growth.

4.1.1 Spatial Ability in Engineering Education

In earlier chapters, we have already delved into the pivotal role played by spatial ability within the context of EE (see Sects. 2.1 and 3.1.1.1). However, recognizing the profound impact of spatial ability is only the first step. As EERs, we bear the responsibility of actively collaborating with educators at various levels to cultivate robust spatial ability among our students. This collaborative effort begins by engaging with teachers at lower grades, where the foundation of spatial understanding is laid.

During the period under study at PoliTo, we had the privilege of engaging in spatial ability training across diverse educational environments. Notably, the SAperI summer school initiative [3] aimed at empowering young girls through direct and intensive spatial training yielded compelling results. Participants demonstrated statistically significant improvements in their spatial ability compared to their peers who did not participate in the program, alongside a remarkable reduction in math anxiety. This reduction in anxiety proved pivotal, as it motivated some of these young students to pursue STEM careers, with the summer school experience serving as a key catalyst for their decisions.

Furthermore, the Matabì project provided a unique platform for integrating various spatial variables. Here, we developed a LEGO Duplo set that facilitated spatial ability training. Moreover, teacher training programs were established, effectively enhancing their spatial reasoning skills, particularly in mental rotation. This, in turn, equipped educators to incorporate indirect spatial training techniques in their classrooms, positively influencing the spatial abilities of young learners.

These initiatives, along with collaborations with institutions like TU Dublin and Dr. Sheryl Sorby's pioneering courses [8, 9], empower students to confidently pursue STEM careers. Integrating spatial training into curricula helps bridge gender gaps and fosters inclusivity in EE.

4.1.2 Addressing Gaps in Prerequisite Knowledge

Addressing prerequisite mismatches, particularly in Mathematics, is essential for student success. Students often face challenges in their early coursework due to gaps in fundamental knowledge.

Recognizing the need for proactive intervention, institutions like PoliTo have implemented innovative measures. The admission test, for instance, serves as a valuable tool in guiding students through a process of self-evaluation. By highlighting specific areas of mathematical deficiency, it empowers students to identify and address these gaps. Importantly, the Test In Laib (TIL) at PoliTo offers students the flexibility to take the assessment multiple times, ensuring a comprehensive evaluation of their readiness.

In tandem, initiatives like the CIAO! (Corso Interattivo di Accompagnamento Online) and !OAIC (Online Accompanying Interactive Course) provide essential support for freshmen in reviewing and reinforcing fundamental mathematical concepts. These gamified approaches not only make the learning process engaging but also bridge the gap in mathematical prerequisites, equipping students with the essential knowledge required for their engineering journey.

These initiatives provide essential support for freshmen in reviewing and reinforcing fundamental mathematical concepts, making the learning process engaging and equipping students with the knowledge required for their engineering journey. EERs play a pivotal role in developing innovative strategies, including gamification, to make learning enjoyable while addressing prerequisite mismatches. Collaboration with teachers at lower grade levels helps create a seamless network of support, ensuring a smoother transition for students as they progress through their educational journey.

4.1.3 The Scientific Study Method in Engineering Studies

The scientific method is foundational in EE. It provides a structured approach to inquiry, experimentation, and problem-solving, equipping students with essential skills. EERs emphasize the practical application of the scientific method through innovative curriculum design and hands-on experiences, such as the "DayByDay" program and the REAL experience. This approach fosters critical thinking and resilience, preparing students for real-world engineering challenges. By introducing students to real-world engineering challenges that demand systematic inquiry, hypothesis formulation, and experimentation, EERs allow students to experience the scientific study method's effectiveness firsthand. This method instills a disciplined and systematic approach to dissecting multifaceted problems, teaching students to embrace failure as an essential component of learning.

In conclusion, EERs significantly influence the educational landscape by integrating spatial ability training, addressing prerequisite mismatches, and promoting the scientific method. Their efforts create an environment where students thrive, develop critical thinking skills, and are well prepared for engineering careers. Through collaboration, innovative strategies, and practical application, EERs empower students to excel and contribute meaningfully to the field of engineering.

4.2 Empowering Engineering Educators: A Cross-Fertilization Action

In this subsection, we explore the complex relationship between teaching and research within the academic landscape of PoliTo. Emphasizing the crucial role of EERs, we discuss how they empower engineering educators to balance teaching and research effectively. Through their initiatives and insights, EERs play a pivotal role in shaping the professional development of lecturers.

4.2.1 The Duality of Teaching and Research

In the landscape of Italian academia, as in many other countries, engineering educators often find themselves navigating the intricate duality of teaching and research. This duality is marked by the simultaneous responsibilities of imparting knowledge to students while engaging in research activities that contribute to the advancement of their respective fields. While these roles are undeniably interconnected, they also present unique challenges that necessitate careful balance and strategic management.

One of the primary challenges that educators face is the significant time investment that teaching demands. Preparing courses, delivering lectures, engaging with students, and assessing their progress all consume a substantial portion of their workdays. These commitments, while crucial for nurturing the next generation of engineers, can sometimes leave limited time and energy for engaging in meaningful research activities.

Additionally, academic roles often come with administrative duties and responsibilities within their institutions. These may include serving on committees, participating in departmental meetings, and taking on various administrative tasks. While these responsibilities are essential for the functioning of academic institutions, they further reduce the available time for research endeavors.

Italian universities typically set high expectations for research productivity among their faculty members. This includes publishing scholarly articles, securing research funding, and actively contributing to academic discourse. Meeting these expectations can be particularly challenging when teaching responsibilities occupy a significant portion of an educator's schedule.

Despite these challenges, educators are committed to delivering high-quality teaching to their students. They continually seek ways to improve their pedagogical approaches, stay updated on the latest educational methods, and adapt to the evolving needs of their diverse student populations. Maintaining this commitment to teaching excellence while juggling research demands adds another layer of complexity to their roles.

This is where EERs enter the picture as invaluable allies. EERs are experts in the field of education, and they bring a wealth of knowledge and resources to support lecturers in managing the duality of teaching and research.

EERs are actively engaged in researching innovative teaching methodologies, technology integration, and curriculum design. Their findings and insights provide educators with practical strategies to enhance the efficiency and effectiveness of their teaching practices. By implementing these research-based approaches, educators can potentially reduce the time demands associated with teaching while delivering an even more impactful educational experience to their students.

Furthermore, EERs offer professional development opportunities tailored to the needs of lecturers. Workshops, seminars, and training programs led by EERs equip educators with the skills and knowledge required to balance their teaching and research responsibilities effectively. Initiatives like the POT [1] and TEACH2TEACH experiences have demonstrated how these programs can empower educators to excel in both domains.

EERs also serve as mentors and collaborators, offering guidance on how to align research interests with teaching activities. This mentorship encourages synergy between the two roles, enabling educators to integrate research into their teaching practices. It promotes a holistic approach to academia, where research informs teaching, and teaching informs research, creating a mutually beneficial relationship as suggested in the Scholarship of Teaching and Learning [7]. Notably, the revision of the Structural Analysis course [2, 4], a collaborative effort between an EER and a lecturer, exemplifies how this synergy optimizes effort and time in course development.

In addition to these roles, EERs play a pivotal role in introducing the research dimension into teaching. In some technical fields, it is increasingly recognized that teaching-related projects can be considered as research endeavors. EERs guide educators in defining these projects as research and assist in the publication process. This approach expands the educator's scholarly portfolio while enriching the educational experience.

Furthermore, the dedication of educators to teaching varies with their career stage. As revealed in semi-epistemological interviews conducted during the OEP investigation [6], career-stage-related factors influence their teaching efforts. For instance, junior educators may focus on refining their teaching methods through iterative improvements, driven by student interactions and outcomes. In contrast, more experienced educators may engage in strategic teaching decisions guided by a deep understanding of their students' needs and broader educational goals.

In conclusion, the duality of teaching and research is a central aspect of the Italian academic context for engineering educators. While it presents significant challenges, EERs play a crucial role in addressing these challenges. Their contributions through pedagogical innovation, professional development, mentorship, and advocacy empower lecturers to navigate this duality effectively. In doing so, educators enrich the educational experience for their students while advancing knowledge in their respective fields, regardless of their career stage.

The comprehensive data presented in Table 4.1 underscores the tangible impact of EER in bridging the gap between teaching and research. With direct interactions with 47 individuals, including a majority of senior-level faculty (22 associate professors and 19 full professors), it becomes evident that the career stage of educators influences their willingness and interest in dedicating time to teaching innovation. Furthermore, the collaboration on 20 publications involving faculty members from 7 out of the 11 departments at PoliTo highlights the extensive reach and influence of EERs in fostering research-integrated teaching practices.

Please consider that some fields have started to witness the emergence of reputable technical journals that prioritize the educational perspective, such as IEEE. This shift further underscores the growing recognition of the importance of educational research within engineering disciplines.

This collaborative effort between educators and EERs fosters a thriving academic ecosystem where teaching and research coexist harmoniously.

4.2.2 Course Approach and Methodologies

The design of courses and the methodologies used for teaching are of paramount importance. These elements not only shape the learning experiences of students but also significantly impact their success in mastering complex engineering concepts. Within this context, EERs serve as guiding lights, illuminating pathways to innovative course design and teaching methodologies that empower both educators and learners.

Instructional system design is the art and science of crafting an educational experience that optimally aligns with learning objectives. It sets the foundation for effective teaching and learning. EERs bring their wealth of expertise to this endeavor, enhancing the overall quality of education.

EERs champion the philosophy of student-centered learning. This transformative approach places students at the heart of the educational process, recognizing their unique needs and the importance of active engagement. Collaborating closely with lecturers, EERs aid in the development of curricula that prioritize student needs, leading to the creation of learning environments that nurture active participation and deep comprehension. At PoliTo, we embrace this philosophy through initiatives like GYM (Grow Your Methodology) where lecturers transition towards student-centered learning.

Active learning has emerged as a powerful pedagogical approach, fostering critical thinking and problem-solving skills. EERs actively research and advocate for the integration of active learning strategies within engineering courses. Together with lecturers, they design activities such as group discussions, problem-solving sessions, and hands-on projects. These activities encourage students to become active participants in their own learning journeys. An illustrative example of this approach is the "We have an IDEA" course, which prompts students to role-play as lecturers and engage in the entire ADDIE instructional design process.

In today's digital age, technology is a valuable tool in education. EERs are at the forefront of researching and recommending effective ways to integrate technology into EE. They guide lecturers in selecting appropriate tools and platforms that enhance the learning experience. From online simulations to virtual laboratories, EER-led research informs the strategic use of technology to facilitate concept comprehension and skill development. As an example, the SMaILE app was developed within the SMaILE project, offering an engaging and interactive platform to teach complex AI algorithms.

EERs place a strong emphasis on formative assessment and timely feedback. They collaborate with lecturers to develop assessment strategies that not only gauge students' understanding but also provide constructive feedback for improvement. This iterative process of assessment and feedback contributes to more effective learning outcomes and helps students reinforce their study method, akin to the "DayByDay" approach.

A fundamental aspect of instructional system design is ensuring that course objectives align seamlessly with desired learning outcomes. EERs work closely with lecturers to define clear and measurable learning objectives. These objectives serve as the foundation for designing assessments and selecting appropriate teaching methods, ensuring that the educational journey is purposeful and coherent.

In essence, EERs serve as educational consultants, infusing evidence-based practices into the heart of course design. Their collaboration with lecturers results in courses that are not only well structured but also finely tuned to cater to the unique needs of students. This, in turn, enhances the quality of EE, creating an environment where students can thrive and reach their full potential.

Engagement is the cornerstone of effective teaching. It is not merely about transmitting knowledge but also about igniting students' curiosity and fostering their active involvement in the learning process. EERs play a pivotal role in inspiring and guiding lecturers to embrace innovative approaches that captivate students' interest.

- **Active learning strategies**: EERs advocate for active learning as a transformative approach. They collaborate with lecturers to incorporate strategies such as peer teaching, problem-based learning, and flipped classrooms. These methods encourage students to actively participate in discussions, collaborate with peers, and apply theoretical knowledge to real-world scenarios. Through this active engagement, students gain a deeper understanding of engineering concepts.
- **Inquiry-based learning**: Inquiry-based learning is a student-centered approach that fuels curiosity and critical thinking. EERs work hand in hand with lecturers to design courses that encourage students to ask questions, investigate problems, and seek solutions. This approach not only enhances students' problem-solving skills but also instills a sense of curiosity and intellectual exploration.
- **Project-based learning**: Project-based learning is a cornerstone of EE. It challenges students to apply their knowledge to real engineering problems, fostering practical, immersive learning experiences. For example, at PoliTo, we have witnessed the transformative impact of project-based learning in the Structural Analysis course. Students work on real-world engineering projects, deepening their understanding and preparing them for future engineering challenges.
- **Multimodal resources**: Diverse learning resources play a vital role in accommodating various learning styles and enhancing comprehension and retention. EERs highlight the importance of incorporating multimedia, visual aids, and interactive materials into courses. Initiatives like CIAO/OAIC! exemplify how these resources make concepts more accessible, leading to improved student outcomes.
- **Community and industry engagement**: Connecting students with the broader engineering community and industry is invaluable. EERs facilitate partnerships between universities, students, and industry professionals. This collaboration includes guest lectures, internships, and industry projects, providing students with practical insights and real-world experiences that bridge the gap between academia and industry. The *Percorso Talenti* initiative serves as a prime example of this industry-academia collaboration.
- **Illustrative case studies**: Illustrative case studies provide students with practical examples that deepen their understanding of engineering principles. EERs introduce real-world engineering challenges, allowing students to apply their knowledge to solve complex problems. These case studies serve as powerful learning tools, preparing students for the rigors of the engineering field. In GYM several proposals were in this direction.

- **Game-based learning**: Game-based learning is an innovative teaching method championed by EERs. It engages students by transforming education into an interactive and immersive adventure. EERs collaborate with lecturers to harness the engaging power of games to enhance the learning experience. For instance, the SMaILE app revolutionizes the teaching of AI by making learning interactive and enjoyable. Through gamified experiences, students not only acquire knowledge but also develop problem-solving and critical thinking skills while having fun.

In conclusion, EERs are catalysts for change, inspiring lecturers to adopt innovative approaches that captivate students' interest and foster active engagement. These approaches breathe life into traditional classrooms, creating dynamic learning environments where students become active participants in their education. By embracing these innovative teaching methods, students not only acquire essential engineering knowledge but also develop critical skills that prepare them for the ever-evolving challenges of the field.

These diverse methodologies, coupled with EER-led initiatives, enrich the educational landscape, empowering students to thrive in their engineering journeys and become the problem solvers and innovators of the future.

4.2.3 Impact Evaluation: Navigating the Complex Terrain of Teaching Research

Teaching is at the heart of EE, and understanding the impact of teaching activities is essential for ensuring students receive the best possible learning experiences. In this subsection, we delve into the world of impact evaluation within teaching research, exploring not only its significance but also the challenges faced and strategies employed in this complex endeavor.

At the core of EE research lies the pursuit of effective teaching methodologies and strategies. Evaluating the impact of teaching activities is not merely an academic exercise; it is a critical process that serves multiple vital purposes.

Impact evaluation acts as a powerful tool for continuous improvement in teaching. By systematically gathering evidence on student performance, engagement, and satisfaction, educators can identify areas for enhancement and refinement. This iterative process enables them to tailor their teaching approaches to better align with learning objectives and improve student outcomes.

Educational institutions face growing demands for accountability, making impact evaluation instrumental in justifying educational investments. It provides concrete data that demonstrate the value and effectiveness of teaching practices, which policymakers, administrators, and stakeholders rely on for informed decision-making regarding resource allocation and educational policies.

Impact evaluation also supports research in EE by allowing educators to investigate the efficacy of various teaching methodologies, interventions, and innovations.

By rigorously examining these elements, researchers contribute valuable knowledge to the broader field of EE, supporting evidence-based decision-making in education.

Conducting research on teaching activities presents intricate challenges, particularly within the academic context. These challenges include resource constraints, ethical considerations, time constraints, and variability in student populations. Resource-intensive designs, such as randomized controlled trials (RCTs), can be logistically challenging and ethically complex. Balancing teaching responsibilities with research activities is demanding, and accounting for diverse student backgrounds adds complexity to research design and analysis.

To overcome these challenges, educators and researchers employ innovative strategies. Quasi-experimental designs, which match treatment and control groups based on relevant characteristics, offer practical alternatives to RCTs. Longitudinal studies, tracking student progress over extended periods, assess the long-term impact of teaching activities. Mixed-methods research combines quantitative and qualitative data collection methods, providing a comprehensive understanding of teaching impact. Collaboration and networking facilitate the sharing of best practices and resources for impact evaluation.

Boyer's model of scholarship [7], which includes the scholarships of discovery, integration, application, and teaching, underscores the multifaceted nature of academic work. Integrating this model into impact evaluation highlights the importance of teaching as a scholarly activity. It encourages educators to view their teaching practices through the lens of research and continuous improvement, aligning with Boyer's vision of a more holistic approach to scholarship.

In summary, assessing the impact of teaching activities in EE is a multifaceted endeavor with significant implications for both educators and researchers. While challenges exist, innovative research designs, collaborative approaches, and a commitment to continuous improvement pave the way for meaningful insights into the effectiveness of teaching practices. The goal is to ensure that engineering students receive high-quality education that prepares them for successful careers in the field.

4.2.4 Global Perspectives in Engineering Education

EE thrives on the exchange of ideas, best practices, and international collaborations1indexInternational collaboration. As we highlight the significance of cross-fertilization in this context, it becomes evident that EERs play a pivotal role in facilitating international connections and fostering global perspectives in Italian academia.

EERs serve as catalysts for faculty mobility and training, enabling Italian lecturers to explore and learn from educational practices around the world. This mobility is not limited to physical relocation but extends to virtual collaborations, exchange programs, and training opportunities.

Faculty networking opportunities allow lecturers to immerse themselves in different educational systems and cultures, gaining fresh insights and perspectives. It opens doors to international collaborations and partnerships, enriching their teaching and research endeavors.

EERs actively promote international training initiatives, encouraging lecturers to participate in workshops, seminars, and conferences abroad. For instance, programs like the school dedicated to young research within the cost action offer excellent opportunities, especially for lecturers in the early stages of their careers, to connect with the EER community and enhance their research skills. These training experiences expose educators to cutting-edge pedagogical methods and the latest advancements in EE, equipping them to bring global best practices back to Italian classrooms.

EE is a global community, and EERs recognize the value of international networking. They actively facilitate opportunities for Italian lecturers to connect with their counterparts worldwide.

EER-led initiatives often involve collaborative projects with international research groups. These collaborations transcend geographical boundaries, fostering the exchange of ideas and expertise. For instance, joint research projects, co-authored publications, and shared resources become avenues through which global perspectives are infused into Italian EE. A good example is the REAL remote laboratory [5] that is used from TTPU students from Tashkent (Uzbekistan) and PoliTo students for a win-win in the economical and resources perspective.

By embracing international experiences and collaborating with educators from diverse backgrounds, Italian lecturers and students gain exposure to a broad spectrum of engineering practices and perspectives. This exposure prepares them to thrive in an increasingly interconnected and multicultural world.

Global perspectives foster adaptability and innovation. They encourage lecturers to explore innovative teaching methods, draw inspiration from global best practices, and adapt their approaches to better serve the needs of a diverse student body.

Additionally, students benefit immensely from a curriculum that incorporates global perspectives. Exposure to different cultures, engineering challenges, and problem-solving approaches equips them with a broader skill set and a more holistic understanding of their field. That is why for the PBL Structural Analysis course [2], we choose as inspirational project a bridge design in Australia and the designer took part at the project assessment.

In conclusion, the role of EERs in promoting cross-fertilization and global perspectives in EE cannot be overstated. They serve as bridges that connect Italian academia with the rich tapestry of EE practices around the world. Through faculty mobility, international networking, and a commitment to global perspectives, EERs contribute to the cultivation of well-prepared, adaptable, and globally aware engineers in Italy.

4.3 Final Recommendations

For educators looking to adopt similar initiatives, several practical recommendations can be made. Start by integrating small-scale pilots with learning innovation design through instructional design to gradually shift towards more student-centered approaches. Utilize open educational resources to enhance teaching materials and reduce preparation time. Engage in professional development programs and peer mentoring to continuously improve teaching practices.

4.3.1 Pedagogical Approaches

This subsection explores various pedagogical approaches in engineering education (EE) research, emphasizing their potential to enrich the learning journey of engineering students. Grounded in extensive research, these recommendations are designed to make significant contributions to EE.

Active learning strategies place students at the center of the learning process, engaging them in activities that promote critical thinking, problem-solving, and practical application of knowledge. Techniques include problem-solving activities, group discussions, hands-on experiments, and interactive in-class activities. These strategies provide a dynamic and immersive learning environment, enhancing student engagement and comprehension.

Project-based learning (PBL) serves as a cornerstone of modern EE, bridging the gap between theory and practice. By working on real projects that simulate professional challenges, students develop practical application skills, teamwork, and critical thinking. This approach extends to challenge-based learning (CBL) and design-based learning (DBL), which emphasize interdisciplinary solutions and iterative design processes, respectively.

Gamification introduces game elements into education, transforming the learning process into an engaging and immersive experience. Key elements include playful learning, rewards and achievements, interactive challenges, and healthy competition. These aspects enhance student motivation and engagement, aligning with our overarching educational goals.

The integration of technology is crucial for pedagogical advancement. Incorporating digital resources such as multimedia presentations, interactive simulations, and online platforms enhances the learning experience. Technology transcends traditional boundaries, enabling virtual laboratories and collaborative online environments that extend the reach of EE beyond the physical classroom. It equips students with essential digital literacy and adaptability for the modern engineering landscape.

Assessment aligned with pedagogy is vital for evaluating the effectiveness of teaching strategies. We advocate for assessments that mirror real-world problem-solving, encouraging students to apply their knowledge and skills. These assessments serve as transformative tools, fostering critical thinking, creativity, and practical application of engineering principles. Aligning assessment with pedagogy bridges the gap between theory and practice, setting higher standards for student achievement and fostering a deep understanding of engineering concepts.

Incorporating these pedagogical approaches into EE promises to transform the learning experience, better preparing students for the challenges of their engineering careers. Each approach offers unique benefits, and their strategic integration creates a holistic and dynamic educational environment that empowers students to excel and innovate in the field of engineering.

4.3.2 Engineering as a Holistic Educational Approach

This subsection advocates for a transformative shift in the perception of EE, viewing it as an ongoing, multidisciplinary journey that begins at an early age. Our recommendations aim to address challenges posed by prerequisite mismatches and enhance foundational STEM abilities, focusing particularly on spatial thinking.

We propose the early introduction of fundamental engineering concepts and problem-solving skills in primary and secondary education. Integrating these principles into the broader curriculum nurtures curiosity and critical thinking from a young age, demystifying engineering and highlighting its everyday relevance.

Central to our vision is the idea of "building step by step" the engineering mindset. Starting with foundational knowledge and progressively advancing to more complex topics ensures that each level of learning contributes organically to the development of engineering skills. This sequential approach fosters a sense of continuity, portraying engineering as an accessible and continuous journey rather than an insurmountable specialization.

Interdisciplinary collaboration among educators across different educational levels is crucial. This collaboration enriches the educational experience by facilitating the exchange of insights, teaching methods, and resources. For example, primary school teachers often have strong pedagogical foundations but limited STEM exposure. Integrating STEM education training during their university studies ensures they are well equipped to teach these subjects, bridging educational gaps and fostering a generation of educators with multidisciplinary competencies.

Addressing prerequisite mismatches is another primary goal. Implementing a structured educational framework provides clarity and guidance, helping students progressively acquire the necessary knowledge and skills for success in STEM. Reducing these gaps creates a more inclusive and equitable educational environment.

Enhancing spatial abilities is also critical. Specific educational interventions and activities designed to improve spatial thinking can significantly boost students' aptitude for engineering and contribute to their overall cognitive development.

This holistic approach to EE emphasizes continuity, inclusivity, and collaboration. By implementing these recommendations, we aim to transform EE into a dynamic and accessible field that empowers students from diverse backgrounds to excel in the evolving landscape of STEM disciplines.

4.3.3 Assessment and Evaluation

This subsection focuses on improving assessment methods to accurately gauge the effectiveness of teaching strategies and enhance the learning process for engineering students. Assessment should not merely measure performance but also guide and improve the educational journey.

We advocate for the seamless integration of formative assessment tools throughout the learning process. These tools act as dynamic checkpoints, allowing both educators and students to continuously monitor progress. Continuous feedback mechanisms empower students to track their learning journey, adapt, and refine their strategies, fostering a deeper understanding of engineering concepts. Assessments should serve a dual purpose—evaluating student performance and promoting learning improvement. By providing constructive feedback, faculty can create a culture of continuous growth and development, motivating students to strive for excellence.

Authentic assessments are essential for effective evaluation. These assessments should mirror real-world engineering challenges, requiring problem-solving, critical thinking, and practical application of knowledge. This approach ensures that assessments align with the complexities of the engineering profession, preparing students for future career challenges.

Technology plays a transformative role in modern assessment methods. Incorporating online assessment platforms, simulations, and digital tools streamlines the evaluation process, creating dynamic, engaging, and efficient methods. Embracing technology enhances the overall evaluation experience for both educators and students.

Continuous improvement is vital for both students and educators. Evaluation from the lecturer's perspective is crucial for refining teaching strategies and course design. Lecturers should use evaluation to:

- Reflect on the effectiveness of their teaching methods
- Solicit feedback from students to gain insights into the learning experience
- Identify areas for enhancement in course design and delivery
- Embrace professional development opportunities to stay current with innovative pedagogical approaches

By adopting a proactive approach to self-evaluation, educators can continually improve their teaching, creating a dynamic and effective learning environment for engineering students.

Through these recommendations, we envision an assessment and evaluation framework that not only measures student performance but also actively contributes to their growth and mastery of engineering concepts. Our approach fosters a learning environment where assessment and evaluation are catalysts for continuous improvement and excellence. By incorporating ongoing faculty development, we emphasize the critical role of professional development for engineering educators. This includes supporting faculty in implementing modern teaching techniques, staying updated with industry trends, and gaining insights from neuroeducation to enhance their teaching effectiveness.

We advocate for lifelong learning for educators, starting from their university studies and continuing throughout their careers. This pursuit of knowledge enables educators to adapt to evolving pedagogical approaches, technologies, and educational theories, emphasizing the value of continuous education and skill development. To create engaging learning experiences, faculty should explore and incorporate modern teaching techniques such as active learning strategies, flipped classrooms, and technology-enhanced instruction, which invigorate the classroom environment and foster deeper understanding of engineering concepts.

Staying informed about industry trends is essential for maintaining relevancy in teaching. Educators should actively engage with industry partners, attend conferences, and participate in professional associations to bring real-world relevance to their instruction. Additionally, neuroeducational insights offer valuable perspectives on optimizing teaching methods, encouraging faculty to explore research on cognitive neuroscience to create more effective learning experiences.

Faculty training programs are crucial in supporting continuous improvement, providing workshops, seminars, and peer mentoring opportunities focused on pedagogical advancements. Recognizing and rewarding teaching excellence within academic institutions is also vital. Establishing teaching awards and honors celebrates outstanding contributions to student learning and inspires educators to strive for excellence. By fostering a culture that values and promotes effective teaching practices, we can create a vibrant and innovative engineering education community.

4.4 Major Findings

Since the inception of this research journey in 2017 as an EER within PoliTo, a series of projects and course revisions have been undertaken. Over these years, active interactions with 47 lecturers spanning across 10 departments and various career stages have taken place. The study highlights the willingness and interest of lecturers to collaborate with an EER, recognizing the crucial role of an education expert in supporting their teaching efforts and fostering improvements.

This finding is particularly significant because it underscores a clear need for educational support and enhancement within the institution. However, despite this evident need, EER is not officially recognized as an established research field in Italy. Career progression in Italy relies on scientific sectors (SSD), and EE is absent from any SSD classification. Consequently, there is no formal acknowledgment or designation for EER within the Italian academic framework.

The major findings of this study can be categorized into three key perspectives:

- **Students' perspective**: The study reveals several crucial insights from the students' viewpoint, such as challenges with prerequisite mismatches, particularly in mathematical concepts and spatial ability. It also emphasizes the need for a robust scientific study methodology in EE and addresses the gender gap in engineering, highlighting the importance of self-perception and role models in driving change.
- **Lecturers' perspective**: The research confirms that within the Italian academic context, education is often seen as a peripheral aspect of a lecturer's role, with minimal impact on career progression. However, it also reveals a general willingness among lecturers to improve their teaching practices, especially when supported in reducing the effort required while increasing the impact of their courses. Key elements for lecturers include cross-fertilization among peers, both nationally and internationally, where EER is more integrated.
- **Technological tole**: Technology plays a crucial role in facilitating the educational journey for both students and lecturers. It is essential, however, to ensure that technology is properly integrated into the learning process and its impact rigorously evaluated.

These findings represent a significant step in recognizing the value of EER within PoliTo and the broader Italian educational landscape. They shed light on the challenges and opportunities in EE, paving the way for future endeavors in enhancing the quality of education and teaching practices within technical universities.

4.4.1 Contributions to the Field and Relevance to Existing Literature

This research makes substantial contributions to the field of EE by shedding light on the challenges faced by students and lecturers within the Italian technical university context. It underscores the critical role of EER in addressing these challenges and fostering improvements. The identification of prerequisite mismatches in both knowledge and skills, along with the emphasis on robust scientific study methodologies, adds depth to the existing body of knowledge in EE. Furthermore, the study highlights the need to address the gender gap in engineering through self-perception and role modeling, which has broader applicability beyond Italy.

The practical implications of this research are manifold. For educators, it emphasizes the value of collaborating with an EER to enhance teaching practices and reduce the effort required. The recognition of lecturers' willingness to improve their teaching, particularly when supported, holds significant implications for institutions aiming to enhance the quality of education. Policymakers can draw from this research to inform decisions regarding support for EE research and the recognition of EER as a vital field within the Italian academic framework. The emphasis on technology as a facilitator in education calls for institutions and educators to invest in the effective integration and evaluation of technological tools.

This research aligns with previous studies in the field, such as the work of Valentine and Williams [10], which highlighted the publishing patterns of engineering academics and the impact on their h-index values. Similarly, this book emphasizes the value of EER in guiding individual engineering educators in planning their research careers, particularly noting that lecturers who actively worked with EER were often in the later stages of their careers, with initial educational publications appearing relatively late compared to counterparts in other countries. These parallels highlight the broader applicability of the findings.

While this research aligns with the recognition of EER's value in EE, it also diverges by identifying specific challenges and contexts within the Italian technical university landscape. The focus on prerequisite mismatches, the gender gap, and Italy's unique institutional context provides a distinctive perspective that complements existing literature. Additionally, the emphasis on collaborative efforts among lecturers nationally and internationally aligns with the global trend of EER collaboration while offering insights specific to Italy.

In summary, this research enriches the field of EE by contributing new knowledge, offering practical guidance for stakeholders, and aligning with broader trends in EER. It serves as a valuable resource for educators, institutions, and policymakers seeking to enhance EE and supports the recognition of EER as an essential component of educational excellence.

4.5 Further Directions

As we move forward, our goal remains to continue the transformative journey of EE, shaping it into a dynamic and responsive field that equips future engineers with the skills, knowledge, and ethical awareness needed to navigate the complexities of our ever-evolving world.

The initiatives described have the potential to be scaled to other institutions and adapted to different educational contexts. Future projects should focus on enhancing engineering education through collaborative efforts, cross-institutional partnerships, and the continuous integration of new technologies. Sharing best practices and lessons learned will be crucial in facilitating this process.

The broader implications for the educational community are substantial. By documenting these initiatives, PoliTo provides a valuable resource for other institutions aiming to innovate and improve their educational practices. The lessons learned and best practices highlighted in this book can inspire further research and development in EE.

Educators, administrators, and policymakers are encouraged to collaborate and continue innovating in EE. Continuous improvement and adaptation to evolving educational needs are essential. By fostering a culture of collaboration and openness, the educational community can drive meaningful change.

Looking ahead, the future of EE in the Italian context is bright, characterized by innovative practices and collaborative efforts. This book envisions a future where both educators and students thrive in an engaging, inclusive, and effective learning environment. The pursuit of excellence in engineering education will pave the way for the next generation of engineers to tackle the world's challenges with confidence and competence.

In conclusion, "Empower Education to Foster Engineering" encapsulates a transformative journey. This book has illuminated the path to empowering education as the cornerstone of engineering excellence. It reveals the critical role of EER in bridging the gap between theory and practice, educator and student, and academia and industry. The findings demonstrate that EE is not merely a specialization but a dynamic and integrated educational journey. By nurturing curiosity, critical thinking, and ethical awareness, we forge a generation of engineers who are not just game players but game changers, ready to tackle complex challenges and shape a brighter future. As we embark on the path ahead, armed with evidence-based insights and a commitment to excellence, we strive to empower education, foster engineering, and create a world where the answer to the question, "Be an Engineer?," is met with resounding enthusiasm—"I'm game!"

References

1. Ballatore MG, Felisatti E, Montanaro L, Tabacco A (2020) TEACH-POT: provide opportunities in teaching. Excell Innovat Teach Learn 5(1):79–91. FrancoAngeli Edizioni. https://doi.org/10.3280/exioa1-2020oa10078. Retrieved from https://www.francoangeli.it/riviste/SchedaRivista.aspx?IDArticolo=66341
2. Ballatore MG, Barpi F, Tabacco A (2020) Work-in-progress: pedestrian bridge application in a fundamentals of structural analysis course inside an architecture bachelor program. In: IEEE Global Engineering Education Conference, EDUCON. IEEE Computer Society, pp 41–45. https://doi.org/10.1109/EDUCON45650.2020.9125140
3. Ballatore MG, Duffy G, Sorby S, Tabacco A (2020) SAperI: approaching gender gap using Spatial Ability training week in high-school context. In: Eighth International Conference on Technological Ecosystems for Enhancing Multiculturality (TEEM'20). ACM, pp 142–148. https://doi.org/10.1145/3434780.3436577. Retrieved from https://dl.acm.org/doi/10.1145/3434780.3436577
4. Ballatore MG, Barpi F, Crocker D, Tabacco A (2021) Pedestrian bridge application in a fundamentals of structural analysis course inside an architecture bachelor program. Int J Eng Pedagogy 11(5):4–18. International Association of Online Engineering (IAOE). Retrieved from https://doi.org/10.3991/ijep.v11i5.20151
5. Ballatore MG, Razza V, Regruto Tomalino D, Stievano IS, Tabacco AM (2022) The educational benefit of a remote automatic control laboratory. A win-win collaboration between Asia and Europe. Educ Knowl Soc 23. Universidad de Salamanca. https://doi.org/10.14201/eks.28495. Retrieved from https://revistas.usal.es/index.php/eks/article/view/28495
6. Ballatore MG, Nascimbeni F, Burgos D, Tabacco A (2023) Multiple roads towards openness: exploring the use of open educational practices within an Italian university. Italian J Educ Technol 31:89–100

7. Boyer EL (1990) Scholarship Reconsidered: Priorities of the Professoriate. ERIC, New York
8. Sorby S (2005) Assessment of a "New and Improved" course for the development of 3-D spatial skills. Eng Design Graph J 69(3):6–13
9. Sorby S, Hamlin A, Veurink N (2008) Impact of changes in spatial skills training. In: Proceedings of the International Conference on Engineering Education (ICEE 08)
10. Valentine A, Williams B (2021) Engineering education and non-education research: a scientometric comparison of 7 countries. In: REES AAEE 2021 Conference: Engineering Education Research Capability Development: Engineering Education Research Capability Development. Engineers Australia Perth, WA, pp 755–764

Index

M. G. Ballatore, *Empower Education to Foster Engineering*,
PoliTO Springer Series, https://doi.org/10.1007/978-3-031-75895-9

The manufacturer's authorised representative in the EU is Springer Nature Customer Service Centre GmbH, Europaplatz 3, 69115 Heidelberg, Germany. If you have any concerns regarding our products, please contact ProductSafety@springernature.com

Printed and bound by CPI Group (UK) Ltd, Croydon, CR0 4YY
30/09/2026
02225607-0010